HF763818

Marine Natural Resources and Technological Development

Routledge Studies in Development and Society

1. Searching for Security
Women's Responses to Economic Transformations
Edited by Isa Baud and Ines Smyth

2. The Life Region
The Social and Cultural Ecology of Sustainable Development
Edited by Per Råberg

3. Dams as Aid
Anne Usher

4. Politics of Development Cooperation
NGOs, Gender and Partnership in Kenya
Lisa Aubrey

5. Psychology of Aid
A Motivational Perspective
Stuart Carr, Eilish McAuliffe and Malcolm MacLachlan

6. Gender, Ethnicity and Place
Women and Identity in Guyana
Linda Peake and D. Alissa Trotz

7. Housing and Finance in Developing Countries
Edited by Kavita Datta and Gareth Jones

8. Peasants and Religion
A Socioeconomic Study of Dios Olivorio and the Palma Sola Religion in the Dominican Republic
Jan Lundius and Mats Lundahl

9. Environment and Development in the Straits of Malacca
Mark Cleary and Goh Kim Chuan

10. Technical Knowledge and Development
Observing Aid Projects and Processes
Thomas Grammig

11. World Bank and Urban Development From Projects to Policy
Edward Ramsamy

12. Poverty Orientated Agricultural and Rural Development
Hartmut Brandt and Uwe Otzen

13. Political Conflict and Development in East Asia and Latin America
Edited by Richard Boyd, Benno Galjart and Tak-Wing Ngo

14. Marine Natural Resources and Technological Development
An Economic Analysis of the Wealth from the Oceans
Marco Colazingari

Marine Natural Resources and Technological Development

An Economic Analysis of the Wealth from the Oceans

Marco Colazingari

Routledge
Taylor & Francis Group
New York London

Routledge
2 Park Square, Milton Park, Abingdon, Oxfordshire OX14 4RN
711 Third Avenue, New York, NY 10017

© 2008 by Taylor & Francis Group, LLC
Routledge is an imprint of Taylor & Francis Group, an Informa business

First issued in paperback 2011

ISBN 978-0-415-95852-3 (hbk)
ISBN 978-0-415-51289-3 (pbk)

No part of this book may be reprinted, reproduced, transmitted, or utilized in any form by any electronic, mechanical, or other means, now known or hereafter invented, including photocopying, microfilming, and recording, or in any information storage or retrieval system, without written permission from the publishers.

Trademark Notice: Product or corporate names may be trademarks or registered trademarks, and are used only for identification and explanation without intent to infringe.

Library of Congress Cataloging-in-Publication Data

Colazingari, Marco.
Marine natural resources and technological development : an economic analysis of the wealth from the oceans / Marco Colazingari.
p. cm. -- (Routledge studies in development and society ; no. 14)
Includes bibliographical references and index.
ISBN 978-0-415-95852-3
1. Marine resources. I. Title.

GC1015.2.C65 2007
333.91'64--dc22 2007000039

Visit the Taylor & Francis Web site at
http://www.taylorandfrancis.com

and the Routledge Web site at
http://www.routledge.com

"I thanked Captain Nemo, and went up to the shelves of the library. Works on science, morals, and literature abounded in every language; but I did not see one single work on political economy; that subject appeared to be strictly proscribed."

Jules Verne, *Twenty Thousand Leagues Under the Sea*,
Hertfordshire: Wordsworth Classics, 1992

Contents

Figures

Tables

Foreword

This book could not have been published at a more appropriate time. With the debate in Europe on the potential of the oceans and seas in relation to the Lisbon Strategy to generate more economic growth and create jobs, and with the launching of the Commission's Green Paper on a Future Maritime Policy for the Union, this book offers a comprehensive coverage of ocean economics and comes as a most inspired contribution to our maritime affairs consultation.

What is of relevance in this book is not only the recognition of the critical role that marine natural resources can play in the world's economy, but the insight into cutting-edge technology for the exploration and exploitation of the ocean's resources, as well as the governance issues raised.

Two characteristics of this work in particular deserve to be underlined: the fact that Marco Colazingari emphasises the dimension and the scale of the issues pertaining to the ocean resources which are global and interrelated; and the fact that he does not limit himself to studying the exploitation of resources which are already being harvested, such as fisheries, but takes a more visionary approach and also includes those resources which hold a great potential to be exploited in the future, and even those which are still to be discovered.

The former falls absolutely within the new holistic and integrated approach that the European Commission decided to adopt in relation to maritime affairs. The latter, in turn, is in line and corresponds clearly to the new all-inclusive approach we are undertaking when thinking on ocean affairs at the European level.

Marco Colazingari's book greatly adds to a new thinking on ocean affairs. This is not a minor achievement as we are today at a crossroads in the way we view oceans and seas management. Currently, we still look at them more on a sectoral basis and, consequently, we may fail to understand fully and to clearly address their problems and the across–the-board impact of our interactions with the seas. Hence, we are unable to take advantage of all opportunities arising from new technologies and scientific knowledge, and of exploiting in full new uses of the seas, such as blue biotech, deeper energy drilling, renewable energies and modern mariculture.

This is what makes the explanations provided in this book so essential to the various relationships and interactions among the physical, chemical, biological, social and economic dynamics occurring in our oceans and seas. Only through such a new approach and by understanding these relationships will we be able to explore, in a sustainable way, our oceans and rise to the pressing challenges ahead, such as the impact of climate change and new weather patterns.

The most revealing aspect of the book, however, is the recognition that, in what concerns ocean affairs, humanity is still very much at the beginning of its adventure. Until relatively recently, lack of proper technology has limited the exploitation of the oceans' and seas' natural resources to fisheries. The technological developments of the recent past allow us to think beyond the traditional uses of our oceans and seas and permit us to plunge into the last frontier of our planet: the three-dimensionality of our oceans and seas.

With Marco Colazingari's book we are sailing through uncharted waters towards this last frontier.

Joe Borg

Member of the European Commission,
responsible for Fisheries and Maritime Affairs.

Acknowledgments

I would like to express my gratitude to all those who have helped me, directly and indirectly, in writing this work. In particular, there are several people that I wish to thank for different reasons.

First, I would like to thank my mentor, Prof. Domenico da Empoli, who—among his numerous intellectual activities—is one of the Italian delegates at the International Seabed Authority. Not only has he guided me throughout my whole academic career but also the idea about writing this book was born during a discussion I had with him three years ago.

Second, I would like to express my gratitude to Prof. Stephen Calleya, my friend and Director at the Mediterranean Academy of Diplomatic Studies (MEDAC), University of Malta. He supported me morally and editorially, always giving sound suggestions and precious advice.

I am also grateful to the EU Commissioner for Fisheries and Maritime Policy, Dr. Joe Borg, for honouring me with writing the foreword of this book.

Encouragement and trust—as well as vital funding—came from the Istituto Diplomatico Mario Toscano at the Italian Ministry of Foreign Affairs, Rome. Support and collaboration to the Italian Chair at MEDAC also came from the Italian Embassy in Malta.

The International Ocean Institute (IOI) of Malta is gratefully acknowledged for the assistance and patience demonstrated during my consultations of their extensive and updated library. In particular I would like to thank Dr. Awni Benham, Dr. Aldo Drago, and Dr. Charles Galdies.

The International Council for the Exploration of the Sea (ICES) of Copenhagen is gratefully acknowledged for its friendly reception and support in providing useful information and food for thought. Special thanks go to the General Secretary Dr. Gerd Hubold and to Jørgen N. Jensen.

My thanks also go to Prof. David Attard, director of the International Maritime Law Institute (IMLI) of Malta for granting me access to the library.

Enduring "daily" support and suggestions also came from my friend and colleague Dr. Omar Grech, Director International and EU Office at the University of Malta.

Mårten Flø Jørgensen is gratefully acknowledged for his technical advice and suggestions on marine biological issues.

A special thank you goes to Ottavio Colazingari, for helping me in drawing graphs with his impressive IT skills.

Strong support, inspiration, and help constantly came at all stages of development of this research from Marit Flø Jørgensen, who also assisted me in proofreading. A warm thank you, Marit.

However, I am the sole person responsible for selections, accuracy, and presentation of information and opinions expressed in this book.

Acronyms and Abbreviations

ADCP	Acoustic Doppler Current Profiler
ARS	Airborne Remote Sensing
AUV	Autonomous Underwater Vehicle
BHA	Bottom Hole Assembly
bm	Billion barrels
BMS	Benthic Multicoring System
BOP	BlowOut Preventer
BT	Bathythermograph
CART	Combined Acoustic/Radio Tag
CBD	Convention on Biological Diversity
CCAMLR	Convention on the Conservation of Antarctic Marine Living Resources
CCD	Carbonate Compensation Depth
CCFZ	Clarion-Clipperton Fracture Zone
CCRF	Code of Conduct for Responsible Fishing
CCSBT	Commission for the Conservation of Southern Blue-fin Tuna
CDOM	Coloured Dissolved Organic Matter
CeDAMar	Census of the Diversity of Abyssal Marine Life
ChEss	Biogeography of Chemosynthetic Ecosystems
CO_2	Carbon Dioxide
COMEX	New York Commodity Exchange
COML	Census of Marine Life
CPUE	Catch Per Unit of Fishing Effort
CRI	Cuttings Re-Injection
CSD	Commission on Sustainable Development
CTD	Conductivity, Temperature and Depth
CZCS	Coastal Zone Colour Scanner
DAC	Diffuse Attenuation Coefficient
Db	Decibels
DGPS	Differential Global Positioning System
DNC	Dissolved Nitrate Concentration
DO	Dissolved Oxygen
DVL	Doppler Velocity Log

EEZ	Exclusive Economic Zone
EMSY	Efficient Maximum Sustainable Yield
ENSO	*El Ninõ*/Southern Oscillation
FAO	Food and Agricultural Organization of the United Nations
FFG	Free Fall Grab
FPSO	Floating Production, Storage and Offloading
GCOS	Global Climate Observing System
GEOSTAR	Geophysical and Oceanographic Station for Abyssal Research
GESAMP	Group of Experts on the Scientific Aspects of Marine Environmental Protection
GMO	Genetically Modified Organism
GOOS	Global Oceanic Observing System
GPA	Global Program of Action for the Protection of the Marine Environment from Land-based Activities
GPS	Global Positioning System
GTOS	Global Terrestrial Observing System
HOV	Human Occupied Vehicle
HPLC	High-Pressure Liquid Chromatography
HROV	Hybrid Remotely-Operated Vehicle
HTS	High-Throughput-Screening
IAS	Invasive Alien Species
ICCAT	International Commission for the Conservation of Atlantic Tunas
ICES	International Council for the Exploration of the Sea
IGOS	Integrated Global Observing Strategy
ILO	International Labor Organization
IMO	International Maritime Organization
INS	Inertial Navigation System
IOC	Intergovernmental Oceanographic Commission
IODP	Integrated Ocean Drilling Program
IPOA	International Plans of Action
IR	InfraRed
IRM	Inspection, Repair and Maintenance
ISA	International Seabed Authority
ITQ	Individual Transferable Catch Quotas
IUCN	International Union for Conservation of Nature and Natural Resources
IUU	Illegal Unreported and Unregulated Fishing
LASER	Light Amplification by Stimulated Emission of Radiation
LIDAR	Light Detection and Ranging
LLS	Laser Line Scan
LME	London Metal Exchange
LWD	Logging While Drilling
MARPOL	International Convention for the Prevention of Pollution from Ships (MARine POLlution)

MDGs	UN Millennium Development Goals
MESSR	Multi-spectral Electronic Self-Scanning Radiometer
mmt	million metric tons
MPA	Marine Protected Area
MRI	Magnetic Resonant Imaging
MS	Mass Spectrometry
MSR	Microwave Scanning Radiometer
MSS	Multi-Spectral Scanner
MSY	Maximum Sustainable Yield
mt	metric tons
MWD	Measurement While Drilling
NACA	Network of Aquaculture Centres in Asia-Pacific
NCI	National Cancer Institute
NMFS	National Marine Fisheries Service
NMR	Nuclear Magnetic Resonance
OBIS	Ocean Bio-geographic Information System
OBM	Oil-Based drilling Mud
ODP	Ocean Drilling Program
OECD	Organization for Economic Co-operation and Development
OMZ	Oxygen Minimum Zone
OPC	Optical Plankton Counter
OSPREY	Ocean Swell Powered Renewable Energy
OWC	Oscillating Water Column
PROD	Portable Remotely Operated Drill
PTO	Power Take-Off
PV	Potential Vorticity
R/V	Research Vessel
RADAR	*R*A*dio* *D*etection *A*nd *R*anging
RFB	Regional Fishery Body
ROV	Remotely Operated Vehicle
rpm	revolutions per minute
SAR	Synthetic Aperture Radar
SBM	Synthetic-Based drilling Mud
SeaWiFS	Sea-viewing Wide-Field-of-View Sensor
SLAM	Simultaneous Localization And Mapping
SLAR	Side Looking Airborne Radar
SONAR	*S*O*und* *N*avigation *A*nd *R*anging
SPM	Suspended Particulate Matter
SRS	Satellite Remote Sensing
SSH	Sea Surface Height
SST	Sea Surface Temperature
TAC	Total Allowable Catches
TAG	Trans-Atlantic Geotraverse
TAPCHAN	Convergent Channel System or Tapered Channel System
TED	Turtle Excluder Device

TM	Thematic Mapper
TS	Target Strength
TW	Terawatt (=10^{12} Watt)
UNCED	United Nations Conference on Environment and Development
UNCLOS	United Nations Convention of the Law of the Sea
UNCTAD	United Nations Conference on Trade and Development
UNDP	United Nations Development Program
UNEP	United Nations Environmental Program
UNESCO	United Nations Educational, Scientific, and Cultural Organization
UNIDO	United Nations Industrial Development Organization
USGS	United States Geological Survey
USV	Unmanned Surface Vehicle
UUV	Unmanned Underwater Vehicle
UV	UltraViolet
VMS	Vessel Monitoring System
VSP	Vertical Seismic Profiling
VTIR	Visible and Thermal Infrared Radiometer
WBM	Water-Based drilling Mud
WEC	Wave Energy Converter
WHOI	Woods Hole Oceanographic Institution
WRI	World Resources Institute
WSSD	World Summit on Sustainable Development
WTO	World Trade Organization

Introduction

This book is about the exploitation of marine natural resources. There are many valuable resources presently extracted from the world's oceans and therefore it is already relevant from an economic perspective. However, the world has a great need to find new resources and the ocean is a huge reservoir of untapped ones. In the future, once the proper technological apparatus is developed, these resources will become available to satisfy the needs of a growing world population.

The objective of this research is threefold:

1. to pinpoint some contentious issues relevant to the oceans' natural resources management and protection;
2. to examine cutting edge technology for exploration and exploitation of the oceans' living and non-living resources (fisheries, bio-products, energy resources, mineral deposits, hydrocarbons); and
3. to detect and identify significant emerging patterns which may determine the development of ocean economics in the future.

To exploit natural resources, two basic conditions are required: 1. Resources must be accessible with the available technology; and 2. the exploitation-commercialization process must be economically viable. This implies that some marine economic activities have to be competitive with the land-based ones.

Humanity has interacted with the marine environment since the earliest times. However, man only very recently began to gain effective working access to the depth of the oceans—intervention being limited to the surface because of the lack of proper technology.

This research focuses on the present importance and the future role of deep-sea natural resources and related technological advances in exploration and exploitation techniques. The exploitation process, which is performed with the required technological apparatus, usually impacts on marine ecosystem and causes environmental damages. Governments usually intervene to regulate economic activities in order to protect the environment. On the other hand, the oceanic ecosystem is characterized by complexity. A complex system is neither completely deterministic nor completely random as it exhibits both characteristics. Numerous interactions among physical, chemical, biological, social, and economic dynamics intersect at many levels. Therefore, many interrelated and multidisciplinary issues arise as soon as a researcher starts to deal with the three-dimensional ocean and the multi-product character of its resources and ecosystem services, which are global in their own nature.

Consequently, as introductory to the core-matter of the economics of natural living and non-living marine resources, a brief overview of the other closely connected issues will be pointed out in order to underline the importance and overall aspects of the challenges.

Since resource scarcity and accumulating pollutants will be the two main global challenges future generations will have to tackle, emphasis will be placed on sustainability and environmental issues. International cooperation is the only solution in trying to solve these worldwide problems.

This study also aims to outline and broadly categorize the range of issues involved in the global marine ecosystem. Given the magnitude of these global and pressing problems, this research may serve as an introduction for further economic studies in this realm of complexity.

For each marine natural resource, this research analyzes:

1. industrial activity and related technological apparatus;
2. environmental impact; and
3. economics and governance.

The first chapter is introductory and serves to outline and define the various subjects of the book. The basic physical aspects and figures of the oceans are highlighted in order to give an idea about the scale and magnitude of the marine ecosystem. The economic importance of marine natural resources and their allocation is stressed, and the contended issue of the economic value of the oceans is examined. The main economic functions of the ocean ecosystem are reviewed and estimated. In addition, the environmental impact of human exploitation is considered, and the concepts of sustainable development and overexploitation discussed. A review of the legal international framework governing the oceans and the institutions created to implement it is also provided. Global governance of the oceans and the problems arising from the inherent common-pool nature of the marine resources are specified, while allocative inefficiencies, international policymaking, and multidisciplinary approach are considered and evaluated.

In the second chapter an outline of the technology used to explore and monitor the marine ecosystem and its natural resources is given. As scientific information-based decisions are generally required to shape sustainable policies, the reader is introduced to marine scientific research, including physical and biological oceanography. The fundamental importance of the technology for accessing the marine environment and for exploiting its resources is stressed. Cutting-edge sensors and platforms presently used in the marine environment, including observations from space, are reviewed and described. Figures and costs of both marine research and its related technological devices are given.

The third chapter deals with the offshore oil and gas industry, highlighting with facts and figures its rising importance in the world's energy market. The reasons for the strong growth in worldwide offshore petroleum production are given, describing the results of equipment innovation and technological progress in drilling units. The main phases of offshore oil and gas industrial activity, from pre-exploration to decommissioning, are described in detail. The chapter also offers a brief overview of the potential of gas hydrates contained in seabed deposits.

The fourth chapter describes the potential of marine-extracted energy in the context of the world electricity market. The worldwide efforts to reduce emissions and to shift the production towards renewable forms of energy are accounted. A review of ocean state-of-the-art energy converters and present Research & Development activity in this field is presented. The functioning of several marine conversion systems, harvesting both thermal and mechanical energy, is explained. Differences and similarities in technology, and economic potential for each exploitation process are also considered. The common features of marine renewable energy systems are highlighted, as are the costs and figures compared with traditional energy sources.

The fifth chapter examines the principal seabed mineralizations and the market potential for their extraction. Genesis and mineral content of a wide range of marine deposits and placers are described. The present status of development of the mining industry and its previous efforts to mine the seabed are reviewed. The hydrothermal vent ecosystem's discovery and main features are reported. The possible environmental impact of mining is assessed, and various mining projects and devices described and valued.

The sixth chapter deals with marine biota and their economic relevance in the world economy. Biochemical and physical factors determining life in the oceans are explained and the technology used to study biota described. The traditional economic theory of fishery and bio-economic models are reported. The growing importance of fish farming is stressed in contrast to the declining supply of capture fisheries, giving the basic facts and figures of the fishery market. Fishery policy and management, as well as subsidies-generated problems, are assessed. Finally, conservation concerns are dealt with, and some recommendations for conservation policy are put forward.

The seventh chapter deals with the biogenetic resources from the sea. An assessment of the growing importance and huge potential of marine-derived products is given. The main features of marine organisms' biochemistry are described and the status of marine biotechnology discussed. The new science called "marine biotic" and cutting-edge pharmaceutical marine research are also reviewed stressing their recent advances and their market potential in future.

1 The Ocean Realm

BASIC PHYSICAL ASPECTS

To have an idea about the magnitude of the problems involved in the analysis of the ocean resources, it is worth recalling the basic physical facts and figures. The world's oceans cover about 71 percent of the planet's surface, occupying an area of approximately 362 million km^2, with an average depth of about 3,800 m and a total volume of 1,370 million km^3 of water. Comparing the volumes of oceans and of dry land, the oceans provide 99 percent of the potential living space on Earth. Two-thirds of the oceans are located in the Southern Hemisphere. Given the oceans' wide dimension, it is necessary to distinguish different scales on which the ocean affects human activities.

On the global scale the oceans play an essential role, performing a number of fundamental ecosystem[1] services. The oceans are not only the ultimate recyclers, but have a crucial function in the carbon cycle, as they are one of the main determinants of the planet's climate[2] and sustain the major part of the planet's biodiversity. It has been assessed (Falkowski et al. 1998) that the oceans absorb carbon dioxide (CO_2) from the atmosphere and release oxygen accounting for about 30–40 percent of the absorption of CO_2 produced worldwide each year.[3] Through the photosynthesis process, microscopic plants called *phytoplankton*[4] dwelling in oceans' surface draw approximately 100 billion tons of CO_2 from the atmosphere and release about 45 billion tons of oxygen annually, equal to 50 percent of the oxygen present in the atmosphere.[5] In brief, without the life-supporting function of the oceans, life on planet Earth simply would not exist.

On a smaller scale, the paramount importance and complexity of the ocean-related issues are evident as well because the natural ecosystems that make life possible on this planet are connected in a dynamic of non-linear processes and interactions at various levels. Non-linearity implies that some or all of the parameters involved in the description of the system are not identical over all magnitudes and regimes of system operations—i.e. parameters change their values with changes in the variables. Given our present scientific knowledge and technological apparatus, this makes the oceanic system very complex, highly indeterminate and definitely unpredictable. Although the oceans' elements and functions are strictly connected in a complex system, it is still possible to highlight many smaller ecosystems included within. Physical, chemical and biological factors greatly vary throughout the oceans and sometimes combine to make distinct marine ecosystems such as coral reefs, kelp forests and hydrothermal vents.[6]

Another consequence of the world oceans' complexity is that, when dealing with their governance, it is impossible to neatly separate naturally occurring ecosystem functions from the effects of the many human activities. Coastal management, for instance, is closely connected to ocean governance, since the coastal zone is the land-sea interface. In addition, both ocean and coastal ecosystems are not static but evolve by natural processes—i.e. plate tectonic dynamics, climatic change, erosion and sedimentation etc. Furthermore, every major human marine activity impacts on every other one and the chain of effects and counter-effects is so integrated that it is not possible to deal properly with these activities separately. The straightforward consequence of such interrelations is the global and interdisciplinary approach needed to deal with oceanic problems.

The ocean is a 3D environment, and the physical elements composing it can be distinguished in (A) vertical and (B) horizontal areas, each of which hosts different types of resources and requires increasingly challenging technology the deeper you go in order to explore the habitats and exploit the resources available.

A. Vertical areas include:
   1. The surface;
   2. The water column, which can be divided into four layers:
      a. The upper level, called *Epipelagic* zone (0–200 m);
      b. An upper-intermediate level called *Mesopelagic* zone (200–1,000 m);
      c. A lower-intermediate level called *Bathypelagic* zone (1,000–4,000 m);
      d. A lower level called *Abyssopelagic* zone (> 4000 m).

Two important specific sub-layers are the Oxygen Minimum Zone (OMZ) and the Carbonate Compensation Depth (CCD). In the OMZ, oxygen content is less than 1 ml/L, with depth of the OMZ layer varying from 100 to 500 m. The CCD is a layer of oxygen maximum, usually located at depths around 4,000 m.

   3. The seabed, the main global feature of which is the Ocean Ridge System, is the great mountain range lying on the ocean floor that encircles the Earth. This immense submarine volcanic mountain chain, around 60,000 km long and, in places, more than 800 km

across, rises an average of about 4,500 m above the sea floor and is the most prominent topographic feature on the surface of our planet.
4. The subsoil.

Each layer is important. The upper layer of the water column, especially the *Euphotic* zone—i.e. the uppermost layer bathed in sunlight where photosynthesis takes place—is relevant because of its high biological activity. This zone, which is the part of the water column that extends from the surface down to the depth where light intensity is reduced to 1 percent of that in the surface (at around 100 m) is the most productive zone in terms of biomass and thus economically significant. *Epipelagic* and *Mesopelagic* zones are important for commercial fisheries, and the seabed host mineral deposits and placers. The subsoil is rich in hydrocarbons and hydrates and is therefore particularly relevant from an economic point of view.

A. Horizontal areas include:
1. The open sea;
2. The continental shelf, which is the submerged extended perimeter of each continent;
3. The coastal area, which is the geographic area along the coast;
4. The near-shore, which is the area adjacent to the land.

Regarding the structure of the horizontal dimension, the beach extends from the shore into the ocean on a continental shelf that gradually descends to a sharp drop, called the continental slope. The continental shelf can be as narrow as 20 km or as wide as 400 km. The water on the continental shelf is usually shallow, rarely more than 150–200 m deep. The continental shelf drops off at the continental slope, ending in abyssal plains (see Figure 1.1).

With reference to the scientific knowledge we posses, although many disciplines interrelate in the study of the ocean, the traditional marine science is usually divided into three main partitions:

1. Physical and chemical Oceanography, which deals with the movements of the water and its chemistry;
2. Biological Oceanography, concerned with all that live in the sea, from the tiny *phytoplankton* to the blue whale;
3. Marine geology, which is the science of the seabed and subsoil.

These disciplines account for our knowledge and understanding of the dynamics that guide living and non-living marine resources, while marine and ocean engineering provides the technological apparatus to access the water environment in which these resources are discovered, analyzed and

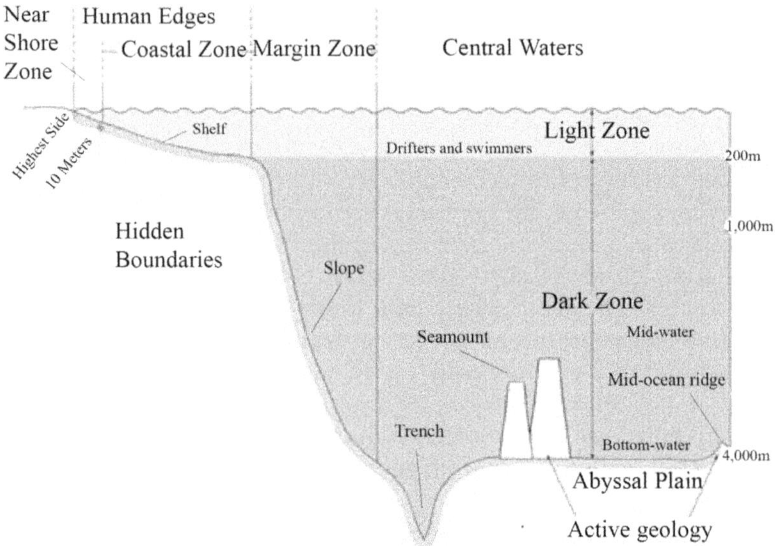

Figure 1.1 Horizontal and Vertical Scales of the Ocean. Based on Census of Marine Life Research Program (2003)

exploited. Therefore, the economic analysis of marine natural resources exploitation must rely on information provided by the above-mentioned disciplines.

ECONOMIC RELEVANCE OF THE OCEAN RESOURCES

In economics, the environment is usually considered as a complex asset providing a variety of services. Apart from its fundamental life-support system function, the environment provides raw materials and energy, which eventually return to the environment as wastes and services. Within these closed-system relationships, living marine resources have always provided relevant economic support for the life of coastal groups, and ocean-produced energetic and food resources influenced social and economic relationships. However, it is only relatively recently that man started to acquire scientific knowledge and technological capability of how to systematically extract resources from the water environment. This is especially true when it comes to concerns relatively to the deep-sea. The oceans are now a complex intersection of many industrial, national and international interests. According to research on the economics of natural resources (Pearce and Turner 1990), the ocean ecosystem supports three main economic functions:

1. Resource supplier;
2. Waste assimilator;
3. Direct source of utility.

Within this general model, the environment and the economic processes are seen as an input-output relationship. The economic activities decrease or add to the stock of renewable resources, reduce the stock of exhaustible resources and generate waste assimilated by the environment. Maximizing economic exploitation can thus degrade the environment, inhibiting its use for other production purposes, or decreasing its direct utility. Excessive wastes depreciate the asset and when the absorptive capacity is exceeded services are reduced. Although it is inherent to models to simplify reality, this basic circular model does not take into account the numerous inter-relations among the various sub-systems composing the environment, nor their inter-dependencies and synergies, which are fundamental to understanding the complex ocean dynamics. It just provides a general trade-off between economic exploitation and the environment's health. The 2005 Millennium Ecosystem Assessment, which is an attempt to establish a global inventory of the planet ecosystem, mentions four categories of ecosystem services that provide economic benefits, each of which can be referred to the provision of marine-related products:

1. Regulating—such as protecting shores from storm surges and waves, maintaining water quality and preventing erosion;
2. Providing—both goods, such as fisheries, minerals, hydrocarbons, energy, bio-products and building materials, and services, such as navigational space;
3. Cultural—such as tourism and spiritual appreciation;
4. Supporting—such as cycling of nutrients, fish nursing and habitats.

Looking at the ocean from an economic perspective, the first thing a researcher would do is try to assess its value in monetary terms. However, assessing the monetary value of the oceans is almost impossible, as traditional economic theory does not provide adequate instruments to measure the real economic value of the oceans. Neither does it provide a method by which to quantify changes in value that result from changes in the condition of ecological systems, i.e. the economic cost of the environmental impacts. Trying to calculate the monetary value of the oceans is therefore meaningless, not only because the measure of such a gigantic asset would be quantitatively imprecise, but also because the pricing of the market system does not work for global ecosystems.

In economics, goods are valuable only in comparative terms and since money has no meaning independent of the system in which prices are determined, it makes no sense to ask how much individuals would pay to retain the world's oceans. The oceans are an essential good, thus meaning a good for which there is no finite compensation for its complete elimination. In giving economic value to the world oceans, it is important to differentiate between: (A) valuations of the ocean ecosystem stock, and (B) estimates of the value of the flow of goods and services from the given stock.

The traditional economic framework is inadequate in valuing holistic concepts such the protection of basic ecosystem functions or biodiversity; however, it can be used to measure *the use value* of ecosystems. While determining the value of the ocean means assessing its value as a stock, determining the value of its resources as a flow is a different concept. This can be economically assessed as the monetary value added by the harvesting production. In this way, economic valuation can be combined with ecosystem functions and related good and service outputs. Nature's service flows can be considered as having a functional economic value.

In other words, from an economic perspective nature can be seen as an asset that provides a flow of goods and services to which monetary valuation can be assigned. The units of service flow may be physical assets, such as fish, hydrocarbons, minerals and genetic resources, biogeochemical processes, such as nutrient throughput and water release, and even cultural-social activities, such as recreational use of the beaches and diving or sailing enjoyment. The marginal unit of the service flow can be valued upon the willingness to pay for their provision or willingness to accept compensation for their loss. Monetising marine-related goods and services sold in the marketplace is a straightforward calculation, as one can look at their market price. On the other hand, to determine the economic value of ecological services, complex valuation techniques must be used.

Unfortunately, ecological valuation techniques such as surrogate market values and survey-based methods are quite abstract and theoretical. However, to put it simply, the value of all goods and services provided by the oceans would give the total economic value for the marine ecosystem. The total economic value of environmental resources is usually divided into three main components: Use Value; Option Value; and Non-Use Value (Tietenberg 2006). The Use Value is the value that comes from the direct use, for example the price of the fish harvested or the minerals extracted. The Option Value is the value placed on the possible future use of the same resources. The Non-Use Value is the existence value that natural resources have even if not exploited. It reflects the human willingness to conserve the environment.

Even though putting a monetary value on the oceans is somehow nonsense, there has been a lot of attention on the economic significance of natural systems (Bockstael et al. 1998). Given the desirability of integrating economic and environmental accounts to have a true global assessment, economists tried to develop methods for valuing non-marketed public goods[7] such as environmental quality, while some ecologists and other natural scientists (Braden and Kolstad 1991; Freeman 1993) have developed their own estimates of the economic value of ecosystem services. Therefore, several methods to estimate natural resources value are available. With direct observation methods, prices are directly observable and calculation possible. When values are not directly observable, surveys to estimate willingness to pay may be used (the so called "contingent valuation"). Alternatively, economists try to infer values through a number of methods such as travel cost value, hedonic property value, hedonic wage value, avoidance expenditure value, etc. (Tietenberg 2006).

Concerning the marine environment, there has been an attempt (Giarini 2004) to value the oceans within the framework of the 'dowry and patrimony' theory, which includes every available resource and asset—material and non-material, monetised and non-monetised— using qualitative indicators and the concept of utilization value. From the traditional economic point of view, the valuation methods used in this attempt are too abstract to be considered quantitatively robust. A well-known but criticized attempt to perform an ecosystem valuation can be found in a paper by Costanza and colleagues (2002) which estimates the annual net worth of the biosphere for the year 2000 at a total of USD 38 trillion, with a value of USD 24 trillion from ocean-related services and a value of USD 14 trillion for what the land offers (see Figure 1.2).

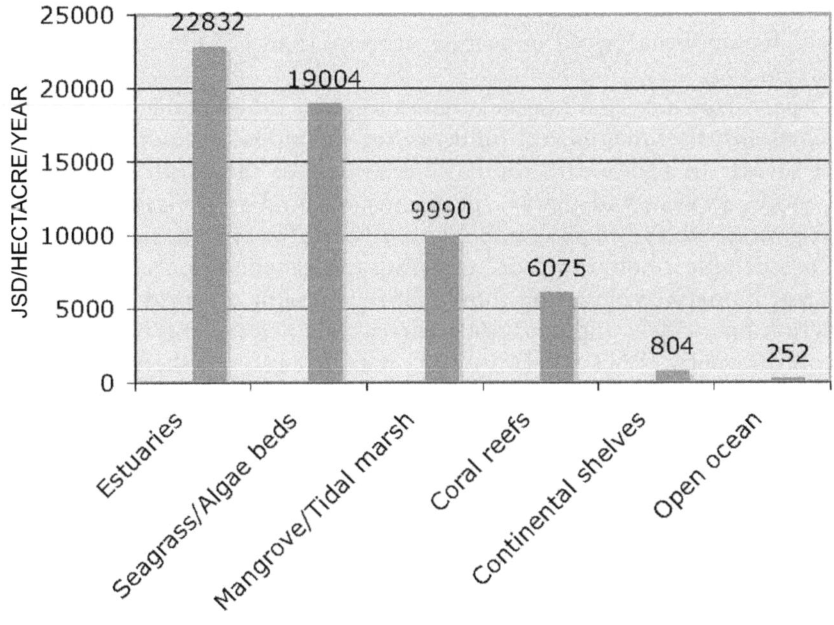

Figure 1.2 Estimated Mean Value of some Marine Biomes, based on Costanza et al. (1998) with permission from Elsevier.

Another recent attempt (UNEP 2006) to assign tangible value to nature's services focuses on two well known marine ecosystems: Coral Reefs and Mangrove Forests. This UN report underlines the vital role of these ecosystems in tourism, coastal erosion and fish nurseries. The value of the coral reefs have been estimated to be between USD 100,000 and USD 600,000 per km^2 a year, while the protection costs of managing marine protected areas amount to USD 775 per km^2 a year. Coral reefs have been priced differently according to the tourist dependence of their location; in Indonesia and the Caribbean for instance, where the net annual benefit from diver tourism amounted to over USD 2 billion in 2000, the coral reefs have been valued at about USD 1 million per km^2 per year. Another typical function of coral reefs is that they can absorb up to 90 percent of waves' force, thus protecting both shore and coastal infrastructures from erosion and damage. One square kilometre of coral reef prevents about 2,000 m^3 of coastal erosion per year. For instance, degradation of coral reefs at Maldives made the local government install artificial breakwaters of concrete, which cost USD 10 million per km^2, restoring over seven years a 250 metre-long eroded beach in Indonesia cost about USD 125,000 per year. The same UN report estimates that coral reef fisheries are worth between USD 15,000 and 150,000 per km^2 per year, including the aquaria fish trade, a high-value, low-volume market in which the price of fish is about USD 500 per kg. Estimates for Europe and North America report that 1.5 to 2 million people have aquaria. Mangroves are even more valued than coral reefs, having been priced in average at more than USD 900,000 per km^2 per year.

Apart from any quantitative attempt to put a value on the ocean and its ecosystems, its fundamental influence on the global economy goes without saying. In addition to the basic relevance of the ocean as a supplier of goods, services and energy, the oceans are used as a relational space in navigation, cables and pipelines. As a *supplier of goods*, the ocean provides a considerable amount of food, minerals and genetic resources. Figures of marine biomass exploitation show that the amount of world fisheries production has widely and constantly increased since the end of World War II until recently (FAO 2004). In 2003 marine capture fisheries and aquaculture produced over 130 million tons of fish food worldwide. Fish is the primary source of animal proteins for 2.6 billion people in the developing countries. It accounts for about 8 percent of animal proteins in Europe and North America, 26 percent in Asia and 17 percent in Africa. Concerning minerals, the ocean's current contribution to the global economy output is providing freshwater by desalinization[8]—more than 3,500 desalination plants are operational in 60 countries—as well as salt,[9] sand and gravel, oil and gas, tin, gold and diamonds. Moreover, its potential as provider of additional minerals in the future is huge, linked to scientific discoveries and technological advances in marine equipments. Another relevant economic potential of the oceans comes from the application of biotechnology to marine biota. Biotechnology is the practical use of living organisms and their genetic content to provide valuable consumption and industrial products. Biomass from the ocean is presently used for applications in pharmaceuticals and beauty products and this market, which is still in its infancy, is expected to grow rapidly over the next decades. According to an estimate (Gorine-Ysern 2003) the budget of biotechnologically marine-derived products accounted for USD 100 billion in 2000. Another estimate for marine-derived pharmaceutical and personal care products reports a USD 400 billion market in 1999 (Newman et al. 2003).

As a *supplier of services* the ocean provides trade routes, tourism and transportation. The ocean is undoubtedly the major natural transport infrastructure worldwide. Even though its relevance for transportation has always been fundamental, the degree of the sea-dependence for merchant transportation has been increasing over the years. Figures for the year 2000 show that, being the cheapest way of moving large quantity of commodities from one place to another, ship transport accounts for 90 percent of the world's trade tonnage (UN 2002). Shipping is a market of more than USD 400 million a year. Tourism, being the largest industry worldwide, was a business of USD 525 billion in 2003 (with 694 million international tourist arrivals), employing 260 million people (11 percent of jobs globally) and the industry is expected to grow considerably in the years to come. According to the World Tourism Organization, tourism receipts represent around 25 percent of total export earnings in the Pacific and over 35 percent for Caribbean Islands. Marine tourism and leisure in particular is a growing industry of more than USD 200 billion a year. Boating alone is a sector valued at more than USD 80 billion per year and cruising is a USD 15 billion sector (Douglas & Westwood 2004).

Concerning *energy resources* from the oceans, there are about 8,000 offshore oil and gas installations operational worldwide, which contribute around 30 percent and 20 percent respectively to the world's energy supply (UN 2002). Moreover, the oceans hold a great economic potential as source of alternative energy, such as tidal power generation, wave kinetic energy and ocean thermal energy conversion. It has been calculated that the daily heat generated by the absorption of solar radiation in tropical seas is equivalent to that of about 170 billion barrels of oil (Busuttil 2004). Renewable energy coming from the ocean is a huge reservoir of power that recently started to be tested for sustainable harvesting. Even though still in its infancy, this sector has great potential and huge investments are presently ongoing to develop a proper technology. When technology will be efficient enough to commercially exploit this source of energy the economic relevance of the ocean will increase logarithmically.

To summarize the economic size of the marine sector, a comprehensive and reliable estimate of the worldwide revenue generated by marine-related industries in 1999 was about USD 1.5 trillion (Douglas-Westwood Associates 2000). The industries included offshore oil and gas (which account

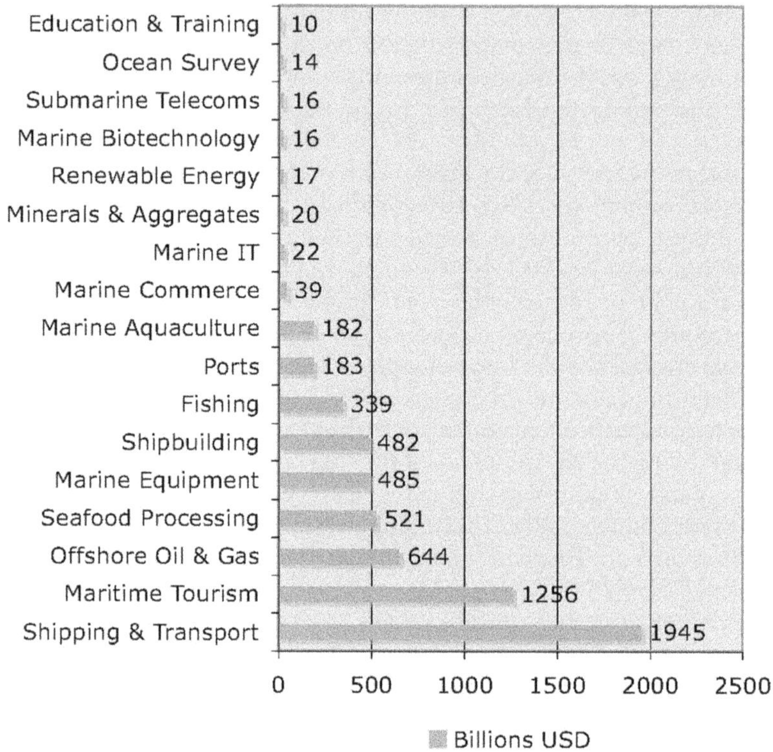

Figure 1.3 World marine industry sub-sectors value, based on Douglas-Westwood Ltd, Opportunities & Challenges for the Marine Industries (2004c)

for about 30 percent of the total), shipping revenues and naval expenditure (more than 40 percent), submarine telecommunication revenue and leisure boat revenue, just to name the top ones on the list. Figure 1.3 and Figure 1.4 show the value of the main marine industries from a more recent publication by the same authors.

Although the world's oceans provide a number of basic ecosystem functions and supply goods and services that in turn generate several marine-related satellite industries, the focus of this economic analysis is on the ocean as a *provider of natural resources*. It is possible to divide the natural resources from the oceans as follows:

1. Biological or living resources, which are in principle renewable;
2. Non-biological or non-living resources, which deplete with exploitation;
3. Bio-power energy resources, renewable.

Renewable resources have a natural replenishment; however, the maintenance and volume of their efficient sustainable flow depend on harvesting behaviour and management. Regarding depletable resources, three concepts are used to classify the stock: Current Reserves, Potential Reserves, and Resource Endowment. Current Reserves are the known resources that can be profitably harvested at current prices. Potential Reserves depend on the willingness to pay and on technological progress—the higher the price, the larger the amount of resources and the more efficient the extraction, the larger the amount of potential resources. The Resource Endowment is the geological concept of natural occurrence of resources. Depletable resources can be either recyclable in various degrees, for examples minerals, or non-recyclable, such as hydrocarbons. Some of these resources are not yet economically exploitable because of the high costs of the technological equipment needed to operate in the harsh marine environment. As a result, some sectors of the present ocean-related industries cannot yet compete with the lower costs of land-based similar commodities but they most likely will in the future.

In our global economy the oceans represent some of the world's most important natural capital for the future. However, the oceans should not be considered anymore as a resilient source of endless resources as they were once believed to be. Their coherent management and protection, based on a sound and continuous monitoring of their conditions, should be of the greatest concern for everyone, as *sinecure* and *laissez-faire* exploitation may upset the delicate balance between available resources and harvesting technology and capacity.

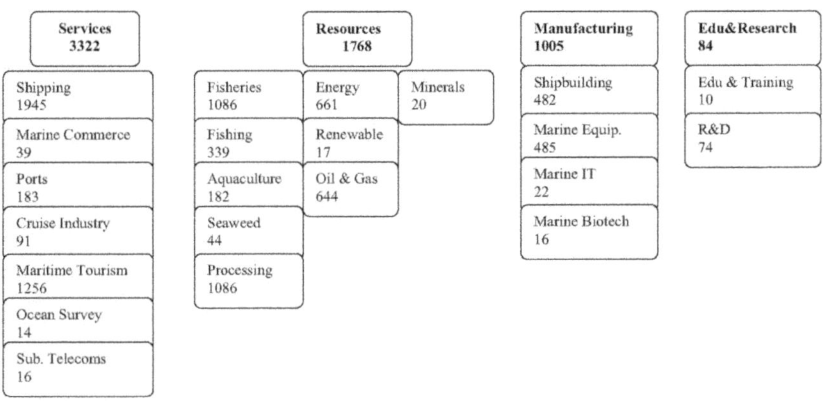

Figure 1.4 World Marine Industry Totals 2005–2009 (USD billion). Based on Marine Foresight Series 1: Marine Industries Global Market Analysis, Douglas-Westwood Ltd, Irish Marine Institute 20

ENVIRONMENTAL IMPACT OF ECONOMIC EXPLOITATION

It is a common view that concern for the environment is a luxury that only the wealthy nations of the world can afford. Only people whose basic needs for food and shelter have been met can then begin to concern themselves with the health of the planet and can permit themselves to be concerned with aesthetics. Therefore, the demand for a clean environment is highly income-elastic. Developing countries have today more pressing concerns, such as fighting poverty and increasing their citizens' per capita income. As a result, environmental-related standards are often considered by developing countries as potential threats to their domestic development or as disguised barriers to trade introduced by developed countries in order to protect their markets. Global public goods are more highly valued by the rich countries of the developed world than by the poor less developed countries, and, therefore, they are marked by strong asymmetric evaluations. As a matter of fact, the asymmetry regarding environmental impact evaluation is not the only one in this field. There is a significant asymmetry also between our understanding of the environment and its dynamics and our capacity to alter it. Our ability to exploit natural resources is far superior to our capacity to control the impact it provokes.

Pollution of the marine environment from human activities is produced by three main sources (Potin 1999):

1. Direct discharge of wastes;
2. Land runoff through rivers;
3. Atmospheric fallout transferred by winds.

The relative contribution of these sources of pollution varies according to:

1. The characteristics of the substances, such as composition, toxicity, volume, persistence etc; and
2. The contingent situation, such as scales of distribution, degree of hazard etc.

The contribution of each source into the total contamination of the marine ecosystem is not a function of a mere addition, since simultaneous impacts of a number of factors can cause synergy effects sometimes unknown. The highest contamination levels of pollution in the marine environment are concentrated in enclosed seas and coastal waters, especially in the Northern hemisphere, being mainly localized on the surface and the seabed. Although there are indeed non-economic aspects of pollution, such as the technological, the political and the ethical ones, from the economic point of view pollution should primarily be considered a problem caused by resources misallocation. From the environmental point of view, the result of the interfacing between the oceans and the land is disastrous, as the ocean is the final big open sewer of almost all human activities. About two-thirds of marine pollution comes from land-based (more than 40 percent) and atmospheric (more than 30 percent) sources, while in the developing countries around 90 percent of sewage disposal is simply dumped untreated into the sea.

Discharges tend to cut across a broad range of economic sectors and usually take place within an area of national sovereignty, where states are reluctant to admit international environmental regulations. One of the main global ocean governance challenges is how to address pollution of the marine environment from land-based activities across different national regimes.[10] Solving the problem of land-based sources of marine pollution is likely to be very expensive, since it requires a revolutionary shift in industrial practices and technology, which is not possible in the short-run. However, the economic dimension of development should integrate the environmental dimension of protection and conservation. The renowned and ever cited concept of 'sustainable development'[11] has in fact the aim to reconcile economic growth with environmental protection requirements. This should be done having also regards for future generations—the so-called "inter-temporal fairness" in resource allocation.

As previously mentioned, most of ocean pollution comes from land-based human activities. The shoreline, which is the area where water and land come into direct contact, has a worldwide length of 620,000 km. Although the oceans comprise about 20 percent of the Earth's surface, more than 50 percent of the world's population lives in coastal areas, i.e. within 100 km of the shoreline, and about two-thirds of all cities over 2.5 million inhabitants are located in coastal areas (World Resources Institute 2001). Due to migration to cities, the mega-cities near the oceans are constantly growing, thus increasing the general trend of moving to coastal areas. UN analysts expect that between 1990 and 2025 the number of people living in urban areas will double to more than 5 billion and that 90 percent of that growth will be in developing countries (Revision Population DB 2004).

It seems inevitable that because of the increasing impact of the demographic growth on coastal areas the clash between mankind and oceans will grow even worse in the years to come. Today the intensive use of the ocean is without precedent, threatening living and non-living resources as never before and adversely impacting—changing or destroying—natural ecosystems. These trends have already gained momentum, and in the next decades we may perhaps succeed in slowing but not in reversing them.

The decline of ocean wildlife is evident in the discouraging figures that describe this trend. The increasing pressure of human overexploitation seriously threatens a huge part of the ocean's living resources. Fisheries should be reduced by 75 percent, as 47 percent of fish stocks are fully exploited and 28 percent are depleted, overexploited or recovering. Fisheries by-catch and discards represent a real shame and amount to about 30 million tons annually (FAO 2001). About 70 percent of marine mammals are presently at risk. Out of 126 species, 88 marine mammals are listed on the International Union for Conservation of Nature and Natural Resources (IUCN) Red List of Threatened Species (UNESCO 2001). In order to help restore depleted fisheries, ecosystems and habitats and to protect the rights of local communities, biological diversity and productivity, since 1972 there have been established about 1,300 marine parks, reserves, refuges and sanctuaries around the world. This means that less than 1 percent of the marine environment is within protected areas, against about 10 percent of the land surface. In addition, many Marine Protected Areas (MPAs) located in developing countries are basically unprotected. Such areas are often referred to as 'paper parks' because of the ineffectiveness of the protection. The World Summit on Sustainable Development held in Johannesburg, South Africa, in 2002, reports that more than 60 percent of the world's coral reefs are endangered. It has been estimated that about 50 percent of the world's mangrove forests have already been lost, while other important sea-grass habitats such as wetlands are being rapidly destroyed (World Resources Institute 2001).

Coastal ecosystems, coral reefs, as well as mangrove swamps constitute the main habitats and breeding grounds for some important marine populations and serve as shields against the power of *tsunamis*. Conversely, it is only by maintaining and improving the environmental quality of the ocean that we can maximize the value of its resources. Consumers are not always aware of the environmental cost of the products they buy. However, in our communication society, information can be almost instantaneous and worldwide spread to a certain extent. Knowledge and public awareness of the importance of these issues could be enhanced. The pro-active participation of all humans is needed in order to integrate the effort towards the implementation of an effective, comprehensive and environmental-friendly approach to global ocean governance. A common concern to each citizen of the world should be to avoid waste of natural resources.

However, the choice facing human society is not between polluted and dirty oceans and clean ones, but rather between various levels of dirt and pollution. Although difficult to quantify, the optimum level of pollution must be determined. This is found where the cost of further reduction in the pollution level begins to exceed the benefit. Estimating costs and benefits of pollution abatement is not an easy task, but it is still possible where economic costs are concerned. The most difficult cost to estimate is the so-called *intangible*, or non-economic cost, such as the value of marine ecosystems (mangroves, coral reefs, reduction of marine life, beach recreation etc). Theoretically the non-economic cost is somewhere between zero and infinite. Therefore, it is subject to under- and over-estimates by different parties and interests. Environmentalists assign very high value to the environment. Some of them would stop any economic activity just because the starfish population in a remote area in the middle of the Pacific Ocean is decreasing, or would prevent shrimp fishing just because a certain number of turtles get trapped in the nets. On the other hand, businesses are less sensitive to environmental issues and assign a low cost to environmental deterioration. Some of them would persist to dump untreated refusal into the oceans just to save the money required for a proper disposal. As every human activity has polluting effects, the main problem is to find the most desirable balance. Economic activity cannot stop just for the sake of conserving the environment. However, at the same time everything possible to reduce the magnitude of the environmental impact must be done. A real breakthrough would be extensive and efficient recycling but this subject is outside the scope of this book.

LEGAL FRAMEWORK OF THE OCEANS

Apart from ancient regulations mainly regarding shipping, transportation and piracy—already present in the Hammurabi Code (2200 BC) and later in Roman law—there have been two main strands of thought in the history of modern ocean law. The first is the Dutch scholar Grotius doctrine of 'freedom of the seas'. This theory was supported by both the faith in the inexhaustibility of ocean resources and by the economic interest of the great naval powers. The doctrine of the father of international law, expressed in the 1609 pamphlet *Mare Liberum*, embodies the fundamental notion of free access to the marine resources that are beyond the jurisdiction and control of individual sovereign states. The English jurist Selden originally pleaded the second line of thought in the 1652 pamphlet *Mare Clausum*. This theory is, on the contrary, the reaffirmation of traditional prerogative on coastal jurisdiction and sovereign state control over the use of marine resources. In the historic development of ocean law Grotius' theory prevailed for centuries until the first UN Convention of the Law of the Sea (UNCLOS).

The international regime of the oceans embodied in the UNCLOS was adopted in Montego Bay, Jamaica, in December 1982 after nine years of complex negotiations. Entered into force in November 1994, the UNCLOS is a mix of the two historical lines of thought of ocean law. On the one hand, the 'common heritage' and the *Area*, embodying both (1) altruistic objectives and concerns toward the less developed countries, and (2) general principles of common human responsibility for conservation and preservation of the ocean resources for the needs of future generations, are conceptual variants of the first notion (*Mare Liberum*).

On the other hand, the concept of Exclusive Economic Zone (EEZ) and the state jurisdiction over the continental shelves are based on notions of state sovereign claims (*Mare Clausum*). Although stated in the 1982 UNCLOS, the present international right of a coastal state to claim sovereignty over its continental shelf is usually identified as originating from US President Harry S. Truman's Proclamation in 1945. This announcement unilaterally extended the US jurisdiction over all natural resources on the nation's continental shelf. The principle was soon followed by other nations and was later laid down in the 1958 Geneva Continental Shelf Convention (UNCLOS I) and reiterated in the 1982 UNCLOS. As a result, nowadays 36 percent of the ocean's volume is under national jurisdiction. This means an area of approximately 37.75 million nautical square miles. Due to the 1982 UNCLOS, the present legal situation of the oceans is now the following:

1. *Territorial Sea*, which is the area within 12 nautical miles from the baseline of the mainland. In this region the coastal state, apart from some exceptions related to navigation, exercises full sovereignty over marine natural resources, both living and non-living;
2. *Contiguous Zone*, which is the area adjacent to the territorial sea with a limit of 24 nautical miles from the baseline. In the contiguous zone the state may exercise control to prevent and punish infringements to its customs, fiscal, immigration or sanitary laws and regulations;
3. *Exclusive Economic Zone*, which is the area beyond and adjacent to the territorial sea. The EEZ may not extend beyond 200 nautical miles from the baseline. The coastal state has sovereign rights to explore and exploit the natural resources contained in the water column, on the seabed and in the subsoil, both living and non-living. However, the coastal state has also duties to conserve and manage the resources;
4. *Continental shelf boundary,* which is the area up to 200 nautical miles from the shore baseline. However, in cases where the continental shelf, i.e. the submerged prolongation of the landmass, prolongs further on, this area can extend up to (A) 350 nautical miles or alternatively, (B) up to 100 nautical miles outward from the 2,500 m isobath.[12] The coastal state exercises sovereign rights to explore and exploit mineral and other non-living natural resources contained therein but cannot affect the legal status of the superjacent waters or the airspace above those waters. Economically relevant is the fact that for the exploitation of mineral resources within its continental shelf, but beyond its EEZ, the coastal state must pay increasing contributions to a supra-national body, the International Seabed Authority[13] (ISA) after the first five years of production, up to a maximum of 7 percent;
5. *High Seas*, which are all parts of the surface and the water column of the ocean not included in the EEZs, in the territorial waters, in the internal waters of a state, or in the archipelagic waters of an archipe-

lagic state. These spaces are open to all states, which have equal rights in navigation, fishing, scientific research, lying submarine cable, over-flying etc;

6. *Deep Sea-beds*, which are the seabed and subsoil extending beyond the marine areas under national jurisdiction or beyond the legal continental shelves. The deep sea-beds make up the so-called *Area*. The *Area*, together with its solid, liquid and gaseous mineral resources at or beneath the seabed, has been declared common heritage of mankind and the ISA is responsible for its management. Through its commercial arm, the *Enterprise*, the ISA can mine the seabed in its own right;
7. *Archipelagic Waters*, which refer to waters containing a large number of scattered islands. Although under national jurisdiction, the archipelagic waters are subject to others' people rights, such as navigation, innocent passage etc.

Figure 1.5 reproduces a draft of the present marine legal zones according to the UNCLOS definitions.

In providing the legal framework[14] governing the use of the ocean and of its resources, the UNCLOS integrates the political, economic, social and environmental aspects of human interactions with the oceans. In a comprehensive perspective on both the marine living and non-living resources, the UNCLOS focuses its attention on the life and environmental dimension of the ocean realm. States have the sovereign right to exploit their natural resources but also a duty to protect and preserve the marine environment. The UNCLOS is considered 'the constitution of the oceans' as it states an internationally agreed framework for the governance and management of the oceans. It is one of the more significant developments in international law since the end of World War II.

The UNCLOS has established other four organisms: the ISA; The Commission on the Limits of the Continental Shelf (CLCS); the International Tribunal for the Law of the Sea; and the Meeting of the State Parties. The

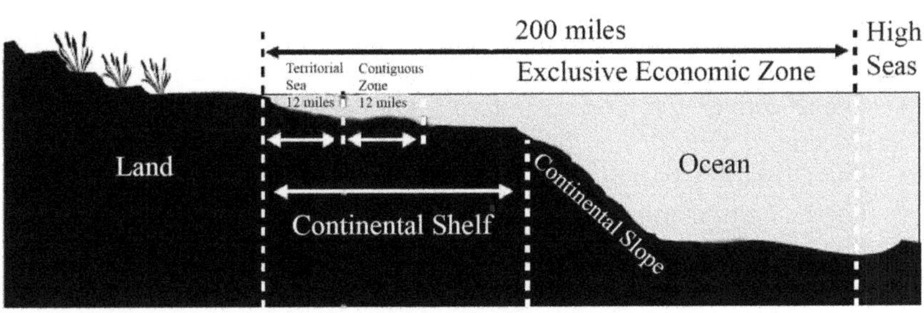

Figure 1.5 Marine Legal Zones, based on information from the Energy Information Administration, US Department of Interior, Minerals Management

most important is the ISA, which is the organization through which states organize and control activities in the *Area*. The ISA is therefore in charge of administering the resources of the *Area*. It came into existence immediately upon the entry into force of UNCLOS and consists of three main organs: The Assembly; the Council; and the Secretariat. The UNCLOS also provides for two implementation agreements for completing the oceans' constitution. The first is the Agreement relating to the Implementation of Part XI of the UNCLOS, adopted in July 1994, which gave birth to the ISA. The second is the Agreement for the Implementation of the Provisions of the UNCLOS relating to the Conservation and Management of Straddling Fish Stocks and Highly Migratory Fish Stocks, adopted in August 1995.

From the economic point of view, in order to ensure the relevant investment decisions of the market forces, the UNCLOS prohibits states to use subsidies, trade barriers and other interventions in the market with the purpose of discriminating between minerals derived from the deep seabed and those derived from land. Moreover, the Implementing Agreement of Part XI of the UNCLOS states that the rates of payment to the ISA by mining companies must avoid artificial measure affecting market competition, such as the creation of competitive advantages with respect to land-based mining. Due to the interrelation of several marine-related topics, apart from the UNCLOS and its two Implementation Agreements, numerous other institutions and international treaties compose the legal framework governing the oceans. The most relevant are:

- Chapter 17 of Agenda 21, which is the document produced in 1992 in Rio de Janeiro, Brazil, by the United Nations Conference on Environment and Development (UNCED). It includes the coastal areas in the 'constitution of the sea';
- The UN Millennium Development Goals (MDGs) for ensuring environmental sustainability;
- The UN 2002 Conference held in March 2002 in Monterrey, Mexico;
- The Johannesburg, South Africa 2002 World Summit on Sustainable Development (WSSD);
- The Global Program of Action for the Protection of the Marine Environment from Land-based Activities (GPA), adopted in 1995;
- The Biodiversity Convention (CBD), which entered into force on December 1993;
- The Climate Change Convention;
- The UN Environmental Program (UNEP);
- The International Labour Organization (ILO);
- The Food and Agricultural Organization (FAO) of the United Nations;
- The Intergovernmental Oceanographic Commission (IOC);
- The International Maritime Organization (IMO);

- The International Convention for the prevention of pollution from ships (MARPOL);
- The UN Conference on Trade and Development (UNCTAD).

In addition, a plethora of regional and sub-regional institutions and agreements are in force at local level. One of the main tasks of international treaties is to ensure that non-compliance can be monitored so that enforcement measures can be taken to ensure observance of the law. Good intentions, commitments and declarations are numerous, rules and regulations abound. However, what should be of concern are also implementation, surveillance and enforcement of the available international governance tools. Otherwise harmful impacts on the marine environment provoked by human activities would continue, especially outside national territorial borders. Ethical and political debates taking place at international summits are essential and enlightening, but what is needed now are practical steps taken at an accelerated pace to reverse the present trend of destroying the ocean's resources.

GOVERNING THE GLOBAL OCEAN

Being the largest ecosystem on the planet and lacking boundaries, the ocean realm inherently needs global governance. National policies exert their effects only within the national boundaries; however, ecosystems do not respect political borders. Therefore, an effective coordination of national and international policies at various levels and scales is needed. It is a common opinion that to effectively achieve global ocean governance societies must find means to agree on decision-making at local, national and international levels. In addition, policy makers and implementing bodies should be supported by a reliable and updated scientific knowledge of ocean processes and sound assessments of marine resources. An appropriate large-scale approach to ocean management has to be built on several international actions stemming from a thorough and well-founded diagnosis of the problems. However, the tremendous amounts of information required for an appropriate management of marine resources is expensive to gather and not easy to interpret. A greater public understanding of the links between ecosystems and human welfare is also needed in order to avoid destructive practices. We already have sufficient and agreed upon scientific knowledge to realize that the ocean ecosystem is so physically and functionally interrelated with the whole planet that effectively governing the global ocean means to govern the planet. Apart from political considerations, the main practical problem in governing the global ocean is twofold:

1. the inherent common and global nature of the oceans;
2. the variety of ocean's resources.

1. The 'Common Heritage of Mankind' as the 1982 UNCLOS established following the speech delivered in 1967 by the Maltese Ambassador to the UN Arvid Pardo, states the oceans represent a common wealth to share not only with all the peoples in the world, but also with future generations. However, viewing the oceans as a common asset implies and brings in management inefficiencies related to public goods (see footnote 5).

As a matter of fact, the oceans are better defined as common-pool resources, which are different to public goods. The efficient allocation of common-pool resources is slightly different from pure public goods. The characteristic of 'difficulty to exclude' multiple individuals from using the resources is shared in both common-pool resources and public goods. The difference is that in public goods there is no rivalry in consumption, while in common-pool resources rivalry in consumption is evident: units of resources appropriated by one party are not available to another. When someone catches a fish, that fish is not available anymore for other fishers. The common and global nature of the oceans is also the arena in which concerns about increasing scarcity of natural resources, pollution problems and the consequent sustainable development issue arise. While pollution problems and the sustainable development approach are quite recent issues, the concern about the increasing scarcity of natural resources goes back about 200 years.

Since the end of the 18th century, there have been two main strains of thought about the future of the world natural resources. The first originates from Reverend Thomas Malthus' well-known essay on the principles of population published in 1798. It argues that the human race will always breed until it exceeds the carrying capacity of the planet and resources will be fully depleted. Competition and market economy will lead to short-run profit maximization and rapid exhaustion of non-renewable resources. The second strain of thought—which is more optimistic—believes that human skills and technological development will solve all obstacles to economic growth and future progress, finding substitutes for depleted resources. This will happen because growth depends more on technical development, education and economic policy than on the stock of available resources. Harold Hotelling, in his study "The Economics of Exhaustible Resources" (1931) argues that in a market economy profit-maximizing mining companies will extract non-renewable resources at the 'socially optimal rate'. Today many economists view physical resources exhaustion as impossible, since rising cost from declining availability of the resources will stimulate substitution. Thus, the efficiency of the market mechanism is believed to be the reason for the non-exhaustion of non-renewable resources.

However, as environmental economists emphasize, it is not possible to rely on the same argument with respect to pollution and many 'common property' renewable resources such as open sea fisheries. From the economic point of view, the most serious environmental problems are in fact related to cases where the market mechanisms do not work. It is for this reason that the greater part of ocean economics is based on a non-property and non-sovereignty reality. Usually people have to pay for the resources they use. Hence, they are led to use only the amount that is worth the cost they have to sustain. Then again, the market price system does not apply to common resources. As the standard theory of collective action supposes, in the absence of a centralized enforcement all resources held in common are subject to overuse, misuse or even abuse. It is also commonly accepted that, unless special measures are taken, un-owned or unmanaged resources are easy prey to 'free-riding'[15] activities. If resources are owned, no matter by whom, it is much less likely that they are going to be over-used or abused. Non-excludable open access systems, of which the ocean is the most fundamental example, are inherently subject to abuse because there is no personal incentive to conserve.

Local and intra-national allocative inefficiencies related to externalities can be corrected by government intervention in two basic ways. The first is to prohibit or regulate abuses by decree. The second is to put a price, charging for the use of the public good and thus incorporate externalities in the normal operation of the price system. This kind of intervention is the so-called 'Pigouvian tax', the implementation of which is far more complex than it may appear. In addition, when the activity needed to enforce a more restricted form of property rights is going to cost more than the benefit of doing so there is no point in government intervention. When transnational externalities, such as the oceans-related ones, are concerned, the only way in which they can be corrected is by the intervention of two or more states. Therefore, they are much more complicated to solve. Given the global common pool nature of the ocean, the related problems involve all or nearly all nations of the world. A nation trying to unilaterally internalize the environmental and social costs of its economic activity will be at a competitive disadvantage to another nation that does not. Its prices will be higher and investment money will flow wherever the highest remuneration is found. This problem implies the adoption of international agreements on a global scale, balancing different and often controversial interests and preferences, reflecting different stages of economic development, cultural matrix, political structure etc. Moreover, as Ronald Coase (1960) has shown, the distribution of property rights[16] needs to be well defined in order to get a Pareto-efficient outcome.[17] This is especially true if a high-bargaining-cost global agreement in different geographic, environmental and socio-economic circumstances is concerned among the 200-some states of the world. Eventually, solutions rest on difficult international negotiations, which are complicated in environmental problems by the different time scale perceptions between cause and effect, both in different countries and among experts. It is common in international environmental negotiations to have calls for different levels of immediacy to deal with and consequently procrastination abounds. Other dimensions involved in environmental negotiations are the degree of uncertainty regarding whether an environmental harmful event will occur and the magnitude and degree of the irreversibility of its effects. Our knowledge of ecosystem dynamics and natural phenomena is still scarce in order to understand all the extents of an environmental impact.

2. The different nature of the oceans' resources has brought into the global arena a plethora of institutions, each of which has different responsibilities and designs its own policy. There are FAO for fisheries, IMO for shipping, UNEP for environment, IOC and UNESCO for marine sciences, UNIDO for marine technologies, UNDP for coastal development, ISA for seabed mining, just to name the principal UN Institutions.[18] As a result, there is a multitude of ocean, environmental and international trade agreements, both at regional and international levels. Since the conclusion of UNCED and the 1997 Special Session of the General Assembly, we have three main international agreements regarding the oceans' sustainable development: The UNCLOS; the Agenda 21; and the Rio Declaration on Environment and Development. Although these institutions and agreements embody common perceptions and goodwill, a wide range of initiatives are dispersed because of the fragmentation of the governance instruments. However, significant progress and institutional changes have indeed been achieved and a number of conventions, agreements and programs of action have been negotiated, adopted and entered into force, to address ocean issues.

The present international basic guideline to the complex management of the planet's resources is the principle of 'sustainability'. This concept was advanced in 1987 by the Report of the World Commission on Environment and Development (Our Common Future). The principle of sustainability basically means that economic development must be pursued in an environmental-friendly way. In this respect, a general equilibrium between the increasing economic exploitation—even more intense year by year because of the pressure due to a growing world population—and environment protection should be found and monitored worldwide. To fulfil this purpose it is of fundamental importance, in order not to deplete the initial stocks of natural resources, to determine the harvesting limits of resource exploitation as function of natural capital growth rates. The interaction between ecosystem's natural capital and sustainable yield must be thoroughly studied in order to forecast economic sustainability (Low et al. 1999).

Both the ecosystem approach and the precautionary approach[19] have been progressively incorporated into international agreements and measures in order to achieve marine protection. Moreover, the 'polluter pays'[20] principle, as well as the principle of preventative action,[21] are now recognized and employed to protect the oceans. However, international legal instruments are generally 'soft law', without real compelling effects and without effective enforcement mechanisms. As a result, in spite of the many legal and programmatic agreements and conventions implemented by the international community in the last years, and the always renewed political and financial commitment at all levels, the oceans still remain the world's dump and their natural resources continue to be overexploited without any regard for marine degradation. Even an inter-agency mechanism specifically dedicated to cooperation among ocean-related international institutions would be inadequate to properly tackle these gigantic problems.

To accomplish a global ecosystem-sound marine governance and management, a worldwide accepted and powerful global agency is needed. Perhaps it is not by chance that the first of the 240-some international environmental treaties signed under the aegis of the UN was the 1946 International Convention for the Regulation of Whales. All countries of the world should agree on effectively empowering an international agency to design global eco-politics and internalize and enforce its decisions at national level. This may sound unrealistic in today's world and actually no one can predict when and which international political processes will compel our planet to establish a mechanism of real global governance. However, this is inevitably the future of the world. No matter which country, global climate phenomena affect every living terrestrial and the Earth ecosystem cannot sustain the present level of unfriendly human pressure.

As already mentioned, human populations are undergoing profound and rapid change in abundance and spatial distribution. According to UN projections, the pace of growth of the world's population is fast enough to push the number of human beings above 9 billion by 2050. Over-exploitation, greed, ignorance and poverty are largely responsible for the degradation of our coasts and oceans. It takes little imagination to visualize a scenario in which the more the population exceeds the carrying capacity of the environment's resources, the worse the human impact and the more individual freedoms must be relinquished (Hardin 1998). Even if in recent years the average rate of population growth has declined, most developing countries still experience substantial increases in their population. There is evidence about the negative effects of population density, especially if coupled with poverty. For, under condition of scarcity, birth control should be one of the major global programs of this new millennium agenda in order to preserve the planet and guarantee the individual freedom of its habitants. The economic approach to population control is indirect and aims at lowering the desired family size considering children as consumer durables. However, the analysis of these theories is outside the extent of this book.

Telecommunication and transportation technologies have already removed physical barriers and travel times are constantly decreasing. Distances are shortened. It is a small world that is becoming smaller everyday. In addition, our planet is already globally connected in a network of relationships and exchanges, both virtual and physical. International trade and mobility volume figures increase steadily each year more than production.

Eventually, human institutions will have to catch up to the global changes and a new international agency in charge of guiding the planet development properly, effectively and environmentally friendly would probably be established. The oceans can be the focal point to set up such an agency.

2 The Exploration of the Oceans

MARINE SCIENTIFIC RESEARCH

Even though more than 1,300 people have successfully climbed Mt. Everest, more than 300 women and men have travelled into space, and 12 men have even walked on the moon, only 2 people[1] have dived to the deepest part of the ocean, spending less than 30 minutes on the bottom (Committee on Exploration of the Seas 2003). Scientific knowledge expands everyday and technology improves—more or less—in accordance to Moore's law.[2] However, human understanding of the dynamics of marine ecosystems is still very limited. The causes of relevant regional and global climatic phenomena are far from being thoroughly understood and consequently it is still impossible to accurately forecast natural catastrophes. Various forms of renewable energies from the ocean cannot be exploited yet because of lack of proper technology. Especially the technological apparatus to explore marine resources of the deepest parts of the ocean is still in its infancy. Oceanographers often argue that we know more about the surface of the moon than about the ocean floor in order to emphasize the need to develop marine research.

The main discipline concerned with marine studies is oceanography. The birth of oceanography as an independent scientific discipline is traditionally attributed to the expedition of the research vessel (R/V) HMS *Challenger*. Along the some 69,000 miles she sailed while circumnavigating the globe from 1872 to 1876, the converted warship *Challenger* drilled more than 650 holes at about 450 sites in all oceans of the world, gathering the earliest extensive data on deep-sea life. The few scientists aboard the ship collected about 13,000 specimens of plants and animals and more than 1,400 water samples, as well as took countless measurements of temperature and salinity and hundreds of seafloor sediment samples. The depths were considered lifeless until a cable with encrusted organisms was recovered from the Tyrrhenian Sea off Sardinia, Italy in 1860. The specimens collected by *Challenger* confirmed that some areas of the deep seafloor host life. Since *Challenger*'s historic expedition, interest in oceanographic research started to intensify. At the beginning of the last century, the steamer *Balke* made the first successful measurement of the speed of the current at a fixed point in the ocean and the International Council for the Exploration of the Sea (ICES) launched an important series of research cruises. This association, actively promoted by the Swede Otto Pettersson, had the aim of combining physical and biological oceanography research. Some years later discoveries made through military research during the two world wars considerably changed the nature of oceanographic research. As a matter of fact, marine technology has frequently originated from military research to later benefit the whole scientific community. During World War I for instance, Sound Orientation Navigation And Ranging (SONAR) was first developed and applied for detecting submarines. Later enhancements created useful applications of this technology for both physical and biological oceanography. In 1920 echo sounder was first used to measure the depth of the ocean. In the 1930s bathythermograph[3] (BT) observations recorded water temperature and depth on a glass slide, providing oceanographers with the first data on the temperature structure of the upper water column.[4] In 1939 a remotely triggered camera sent back the first image of the seabed. Since these main historical breakthroughs of marine technology our insight into ocean processes and dynamics has increased dramatically. During the Cold War, billions of dollars were invested in marine research for military intelligence purposes, developing more sophisticated systems to collect information underwater. The hydrophone for instance—which is an underwater microphone developed by military intelligence research and capable of discriminating sounds in the water[5]—was lately converted to civilian use and became part of the instrumentation for scientific research. Physical, chemical and biological oceanographic research has traditionally relied on lab analysis of water and on microscope examinations of living specimens.

Today marine research and exploration typically employ optic and acoustic sensor technology and modern research facilities and infrastructure. The relatively recent possibility to obtain information remotely from a number of different platforms has increased the quality and the quantity of the data. Data is processed electronically with a tremendous saving of time and increased reliability. Even though well-equipped R/Vs are still the principal research platform, marine scientists acquire data from sensors located on satellites (Satellite Remote Sensing, SRS), aircrafts (Airborne Remote Sensing, ARS), robotic underwater vehicles, coring and drilling devices, moorings and buoys etc. As a result, the previous informational gap due to the limits of *in situ* measurements is shrinking and global-scale vision is expanding. Oceanographic research instrumentation, with the assistance of high-density data processing systems, definitely improved the accuracy of ocean scientific surveying, thus providing a more reliable and detailed information set.

There are a number of sound reasons to promote and support worldwide exploration and scientific research of the oceans. From a planet-system perspective, the ocean is the largest ecosystem and therefore understanding the functioning of its dynamics is of fundamental importance if we are to apply scientific knowledge to manage the planet and exploit its natural resources in an efficient and sustainable way. Marine ecosystems are increasingly threatened by a variety of human and natural impacts. Hence, the more we understand about global marine processes and linkages, the easier we can monitor[6] and predict the environmental effects of our activities and distinguish them from changes caused by natural impacts. At present, it is impossible to predict and manage effectively the impact of our activities—especially in the long run—because the quality of information on many ocean and global processes is still too poor. Consequently, ocean governance is still heavily characterized by decisions in conditions of uncertainty.

The urgent need to properly tackle the potential impact of global warming and other climate-related phenomena such as *El Ninõ*/Southern Oscillation (ENSO) and *tsunamis* have largely contributed to the development of marine environmental activities, including monitoring, sensing and modelling of climate and water quality. Following the early interest in environmental issues during the 1970s, the marine environmental market has grown fast during the last decades, widely supporting innovative research and education in order to enhance ocean and coastal ecosystem's stewardship. Marine scientific observation and measurement, therefore, have an important role to play in environmental management and conservation. To fully understand fundamental oceanic processes and key ecosystem variables, physical and chemical oceanography need to acquire long time-series data. There are presently good examples of four-dimension (the three spatial dimension plus time) data gathering, such as tide gauge stations, the routine collection of temperature data by commercial ships and some local physical and biological measurements. Best practices such as the Continuous Plankton Recorder Surveys in the North Atlantic, which began in 1931, or the more recent Global Oceanic Observing System (GOOS) are examples of paramount importance for improving our knowledge on ocean ecosystem dynamics and global changes.

On a smaller scale, reliable and widespread scientific information would assist in managing effectively and efficiently local marine resources and coastal areas. Regarding governance, scientific information-based decisions would contribute to shape appropriate and sustainable industrial policies at regional and local levels. Regarding businesses, scientific knowledge would provide valuable information to a number of marine stakeholders. Decisions on capture fisheries depend on knowledge of fishing stocks, their distribution and life histories. Additional information about ocean circulation patterns, chemistry and seafloor would improve capture fisheries management and pave the way for the adoption of sustainable practices in this sector. Scientific understanding of all aspects of aquaculture and its environmental impact is essential as well, since fish farming is the sole alternative way of meeting the increasing world demand for seafood, which will soon exceed the decreasing supply of wild stocks. A more detailed ocean mapping and charting would serve numerous sectors of the marine industry, including offshore oil and gas companies, prospectors looking for minerals on the seabed, navigation liners, pipelines and cable placing companies and many others, such as for instance the nascent sector of shipwreck salvaging.

Another reason to promote marine scientific research and exploration is that the ocean is a huge source of renewable energy and still contains unknown quantities of biological and mineral resources.[7] This is confirmed each time new systematic marine research is undertaken. Biologic explorations in even a small portion of the ocean always provide a rich collection of newly discovered organisms. Mineral prospecting and exploration always bear hope for new bonanzas to be found at sea. Offshore oil and gas reservoirs continue to be uncovered and oil corporations are reaching and producing at depths once thought unimaginable. As financial and environmental factors will make land-based mining and hydrocarbons extraction more and more difficult, government and private companies would necessarily begin to look for the resources at sea with increasing interest. Moreover, vast portions of the ocean have not been systematically examined yet and consequently their geochemical and biological characteristics still remain indefinite. This is particularly true for the oceans in the Southern hemisphere—which are the largest—and the deep-sea. Estimates suggest that about 95 percent of the oceans and 99 percent of the oceans' floor are still unexplored (NOAA 2000). Discoveries made in recent years offer exciting scientific perspectives and economic opportunities for mankind. In 1977 for instance, while diving with a research submersible at a water depth of 2,400 m, new organisms were discovered at the Galapagos Rift hydrothermal vent field by an oceanographer[8] conducting the first photographic survey of the region (Weiss et al. 1977). Surprisingly, new species and organisms such as *Thermophyles* and *Extremophyles* can be found in the oceans.[9] The discovery of hydrothermal vents, cold seeps, gas hydrates and giant coral mounds in the last years indicate high probability of discovering novel marine habitats and fauna elsewhere. Exciting discoveries are made in the oceans almost every year. In 2005 a new community of soft-walled, single-celled organisms named *Foraminifera* was discovered at Challenge Deep, which is the lowest part of Mariana Trench, the ocean's deepest point.

Since no nation owns the high seas—which are the common heritage of all mankind—and no nation alone has the intellectual, financial and technological capability to undertake a thorough program of global ocean exploration, it is a common and agreed opinion that the exploration of the ocean requires sustained efforts with wide international participation. As a matter of fact, ocean research has a tradition of international cooperation and partnership, necessitated by the nature and vastness of the oceanic environment. Moreover, since marine ecosystem dynamics are the result of a combination of physical, biological, and chemical phenomena cooperation, coordination and integration of marine sciences should be enhanced through trans-disciplinary international teams, at least in the most relevant thematic areas. Good examples of international and inter-sector partnerships in the field of marine science are numerous, however, this trend should continue to be stimulated, developed, and widely supported. The U.S. Committee on Exploration of the Seas (2003) estimates that to undertake a truly large-scale ocean exploration program would require approximately USD 270 million for first-year funding. Thereafter, annual operating costs would "decrease" to about USD 110 million. Even though the largest amount of marine resources is located within the continental margins—and is therefore part of coastal states' EEZ—a real global, well-planned and interdisciplinary program of ocean exploration could lead to unexpected discoveries of the potential of some of its living resources and the amount of its nonliving resources. It is difficult to predict what discoveries are still to come from the oceans. However, it is obvious that ocean exploration would improve the accuracy of our predictions of global climate change, discover novel resources to produce new products, inform policy choices with better scientific data and understanding, and allow a better stewardship of the ocean and a more environmentally friendly exploitation of its resources.

MARINE SENSOR TECHNOLOGY

Oceanographic research and marine economic activities have a strong direct relationship with marine technology, since marine technology provides the technical support for all marine applications. Marine technology is the discipline in charge of the design and implementation of equipment, methods and procedures for all kinds of operations in the marine environment. It provides ocean structures and platforms, water and underwater vehicles and robots, aquatic sensors and sampling systems, communications and electronic equipment. Marine technology requires a deep understanding of how physical and chemical laws interact in the ocean to create such a unique environment. In implementing its product, marine technology and engineering has to take into account hydraulics, wave movements, wind conditions, currents, ice formation, salt and corrosion, huge pressure etc. The main practical problem posed by the marine environment is that water is much denser than air, about 1,000 times. Hence, it is much more difficult to access and operate. The harsh, turbulent and corrosive conditions at sea and the huge pressure at depth make the technology required for marine operations tricky and expensive. However, today engineers can devise more efficient ways to deal with sea corrosiveness and its crushing pressure because of the availability of new high-tech materials such as Teflon-coated titanium, ragged glass spheres and dense plastic composites (Irion 1988). The deepwater sector technology—which is at the forefront of the marine industry—is expected to continue its growing trend and to reach annual investment expenses of over USD 20 billion by 2010. The demand for deepwater equipments is forecasted to increase at an annual growth rate of more than 7 percent between 2005 and 2010, with particularly strong growth coming from the Asia and Latin America regions. The 'Golden Triangle' of deepwater oil and gas—namely Africa, the Gulf of Mexico and the Brazilian areas—will still account for some 80 percent of global deepwater expenditure, thus confirming oil and gas technology as the locomotive of the entire marine civilian industry (Douglas-Westwood 2005). Most underwater technological innovations in the marine research and in other commercial marine sectors spin off from the offshore oil and gas sector, which is the richest and most advanced marine sector.

Another growing deep-sea sector is cabling. Due to the increasing demand in the international TLC market—mainly generated by the widespread use of Internet and corporate data exchange, the submarine cable industry has been expanding quickly in the last decade. To satisfy the overwhelming demand, the cable companies are working at full capacity to install new fiber optic cables in the deep sea and a new generation of advanced cable trenching equipment is underway to respond to this requirement. Water and underwater marine technology finds civilian applications in both scientific research and economic activities related or located in the marine environment. Often the same equipment and methods are used in both fields, however, during the production phase, commercial operators must employ particular devices that are tailored for each specific activity. Since these specific devices differ according to the type of goods harvested, they are dealt with separately discussing each marine natural resource. The subsequent sections of this chapter, therefore, illustrate only the shared technological apparatus that is commonly employed in marine scientific research and in most marine commercial operations. To analyze marine equipment and methods it is useful to separate (A) Sensor technology from (B) Platform technology in which the sensors are positioned.

Sensor Technology

Marine sensors are basically of two types: optic and acoustic. Acoustic and optic methods represent two of the main tools for marine research and are used for observation and measurement at various scales. Measurements can take place *in situ* or remote. Remote sensing in marine science usually refers to measurement of a physical, chemical or biological parameter from satellites or aircrafts. However, in general remote sensing means the acquisition of information from a distance—i.e. without being in direct physical contact with the target. Matter emits or reflects different forms of energy. Sensors are electronic devices able to detect and measure physical forms of energy such as electromagnetic radiations (**optic** sensors) or sound waves (**acoustic** sensors) emitted or reflected by matter. As previously stated, remote sensing systems and techniques—both optic and acoustic—were originally developed for military intelligence use. However, some years later they were declassified and became available for civilian applications. Although improved and refined, the basic technology behind these devices has remained relatively unchanged in recent years, the major innovations being digital processing and post-processing activities. Current data analysis techniques provide software for geometric correction, image enhancement and data mining—i.e. the process of automatically searching large volumes of database information on the basis of association rules. As a result of these improvements, the recorded signal is more precise and easy to interpret, since it is usually visualized on high-resolution computer screens.

In marine science remote sensing is of extreme importance in order to collect data, detect trends and understand various phenomena. While chemical and biological parameters are more difficult to measure, remote sensing technique is especially useful to collect physical data of the ocean. Optic remote sensing technology is very useful to measure wide-scale "surface" phenomena. However, since electromagnetic radiations do not penetrate water more then few meters, optic sensors are generally unsuitable for underwater applications. On the other hand, sound waves travel farther and faster underwater than in air and, therefore, acoustic technology is usually more appropriate to gather underwater information.

Optic Sensor Technology

Electromagnetic radiations, or photons, are part of a continuous spectrum of light the wavelength of which—from the shorter to the longer—includes gamma, X-rays, ultraviolet (UV), visible light, infrared (IR), microwaves and radio waves. Wavelength portions of interest in remote sensing are UV, visible light, IR and microwaves. Electromagnetic radiations undergo a process of interaction with the atmosphere that depends on their wavelength and on the distance travelled. This process, called scattering, results in a redirection of the waves from their original course. The longer the wave's frequency and the shorter the distance from the target, the less biased the detected signal will be. Each optic sensor has a defined spectral resolution—i.e. it can detect the electromagnetic backscatter radiations of the target only over a limited range of wavelengths. Multi-band sensor systems are made of a number of detectors, each of which is sensitive to a different spectral band. Multi-band sensor systems such as the Multi-Spectral Scanner (MSS) and the Thematic Mapper (TM)—employed onboard the satellite platform Landsat—have respectively 4 and 7 bands of different spectral detection, with 6 and 16 detectors for each band. Advanced multi-spectral sensors—called hyper-spectral sensors—can detect hundreds of very narrow spectral bands throughout the visible, near-infrared and mid-infrared portions of the electromagnetic spectrum. Usually electronic sensors acquire data by mechanically sweeping their field of view either along or across their track. This is called a scanning system. The portion that the sensor scans on each side of its sweeping axis is called "swath".

Optic sensors can be either passive or active.

A **passive sensor** (or radiometer) measures radiations that are naturally produced by solar energy, either reflected or absorbed and re-emitted by the target. A passive optic sensor useful for marine sciences is the thermal sensor, which can sense thermal radiations in the upper part of the IR range of the electromagnetic wavelength, allowing the detection of the target's temperature. Apart from thermal detection—where the radiation is emitted by the target and not just reflected—a passive sensor works only during daylight. This, of course, represents a remarkable limit of passive sensor techniques.

An **active sensor** provides its own energy radiation, which is first transmitted to the target, which reflects the signal, and then received back and measured. Therefore, an active sensor works even without daylight and can be employed also at night.

Regarding marine applications of these technologies, the optical properties of the surface can be determined by the presence of dissolved and suspended matter. Under normal meteorological conditions visible light penetrates marine water to a depth of tens of meters, however, when constituents' concentration increases and water becomes more turbid, light can penetrate the water column less. Different characteristics of water constituents have different spectral signatures. Optic sensors, detecting differences in the spectral signature, give information about the characteristics of the water.

The most common active sensor system employing optic technology is RADAR.

RADAR (RAdio Detection And Ranging) detects objects and determines their position (range) by transmitting and receiving short pulses of microwaves from an antenna and recording the strength and origin of the reflection waves. The time necessary for the energy to travel to the target and return back to the sensor determines the distance of the target. The resolving power of the radar—i.e. its capacity to differentiate between targets—is determined by the wavelength of transmission. Microwaves are transparent to the atmosphere and consequently RADAR can acquire data regardless of light and weather conditions. RADAR can be either: imaging or non-imaging.

An imaging RADAR displays the backscatter characteristics of the detected object in the form of a picture on a screen. An example of imaging RADAR is the Side Looking Airborne Radar (SLAR), which is carried on aircrafts. This sensor scans an area not directly below the aircraft but at an angle to the vertical, hence the term "Side Looking". A further development of this kind of RADAR is the Synthetic Aperture Radar (SAR), in which the resolution is independent from the target distance and therefore remains constant across the entire swath. However, imaging RADAR has few applications in oceanography.

Non-imaging RADAR records only one specific physical parameter. Examples of non-imaging RADAR include the RADAR scatterometer and the RADAR altimeter. A RADAR scatterometer has several applications in marine science, including the measurement of the roughness of the sea surface, the wind speed and iceberg spotting. RADAR altimeter has marine applications for height measurement of the sea surface.

LIDAR (LIght Detection And Ranging) is another optic device that finds numerous applications in marine science. It is an active sensor that emits and receives light in both the visible and near-IR wavelengths. LIDAR employs LASER (Light Amplification by Stimulated Emission of Radiation) technology to generate short, high power light pulses. Put it simply, LASER is a device for producing light (photons) by emission of energy stored in a molecular or atomic system. As the light pulse passes through the atmosphere, the backscatter light is detected by an optical sensor receiver and is electronically analyzed to provide a measurement of its intensity. Two kinds of LIDAR have applications in oceanographic studies: bathymetric LIDAR and fluorescence LIDAR. They both require cloud free, no-haze no-fog conditions to operate.

Bathymetric LIDAR generates simultaneously a blue-green signal and a near-IR signal. The near-IR signal does not penetrate the water and is directly reflected by the sea surface and recorded by the sensor. The blue-green signal penetrates the water and is reflected by the bottom before being recorded by the sensor.

Fluorescence LIDAR, or fluorometer, is a sensor that measures and records the emitted fluorescent light induced by the interaction of the transmitted blue-green light with the target. Fluorescence is an optical phenomenon that occurs when the light-strike absorbed by matter provokes a molecular excitation that causes the material to re-emit light at a different wavelength. This sensor has applications for ocean science such as chlorophyll mapping and pollutant detection, especially for spotting naturally occurring or accidental oil spills.

Acoustic Sensor Technology

Acoustics is the study of how sound waves are created, transmitted and received. Underwater acoustics studies the propagation of sound underwater. Sound waves have a speed propagation of about 1,500 m/s underwater (five times more than in air) and a lower attenuation within the distance.

The most important instrument in underwater acoustics is **SONAR** (Sound Orientation Navigation And Ranging) or echo sounder. SONAR was originally developed to detect submarines, however, it soon began to have applications also in civilian navigation. SONAR can be either: passive or active.

Passive SONAR is just a hydrophone that simply listens for sounds emitted underwater. The **hydrophone** is the basic sensor of underwater acoustics. Hydrophones are microphones that transform sound waves propagating underwater into electric signals. Amplifying the electric signal recorded by the hydrophone, it is possible to measure sound waves accurately. A single hydrophone records sounds arriving from any direction and covers a wide range of frequencies. In marine applications, arrays of several hydrophones are usually employed in order to obtain a directional and more sensitive detection apparatus. Hydrophones commonly used in marine science are of three main types: portable, stationary and towed.

A portable hydrophone system consists of a single hydrophone and a waterproof pressure case that contains the electronics (mainly the power and a recorder) required for the system to operate autonomously for extended periods of time. They can be deployed on the sea bottom or even attached to animals. The major disadvantage of portable hydrophones is that data are only available once the instrument is recovered. Stationary or permanent hydrophone systems consist of an installation of hydrophone array connected through an underwater communication cable transmitting the acoustic data of the selected area in real time. Towed hydrophonic arrays are pulled behind a R/V and continuously detect sounds along the route during navigation.

Active SONAR and echo sounder emit pings—i.e. pulses of sound—towards an underwater target and then receive and record the reflected echo. Even if they are basically similar, the main difference between echo sounder and SONAR is that echo sounder transmits the beam of sound to the target in a vertical path, while the SONAR's beam is mainly horizontal, even if can be oriented. The principal components of echo sounder and SONAR are the transmitter, the transducers, the receiver and the display unit. The transmitter produces pulses of electric oscillations. A transducer converts this electric energy into sound energy and another transducer converts back the sound waves of the returning echoes into electric energy. The receiver amplifies the electric signal produced by the second transducer so that it can be recorded or displayed in a screen. SONAR can transmit in different ways or systems. A single-beam system only measures the distance of a target from the receiver. A dual-beam system records the echoes of two concentric conical beams of different wavelengths at the same time, while a split-beam system measures acoustic signals at four independent receivers.

The most advanced type of SONARs are the side-scan SONAR and the multi-frequency echo sounder (also called multi-beam SONAR). **Side-scan SONAR** transmits and receives sound waves by sweeping its field of view on each side of the track. **Multi-frequency (MF) echo sounder** is composed by several acoustic sensors, which allow recording of echoes at different frequencies at the same time. These systems permit a detailed mapping of the seabed and are capable of discriminating among different objects or species of fish and have therefore numerous applications in marine research. Multi-beam SONAR has the capability to scan in a single ping a wide angular aperture (90, 120 and even 180 degree field of view) with high resolution (up to 128 beams) enabling complete coverage of the water column to horizontal ranges up to 500 m. Multi-beam SONAR can easily be converted into echo sounder with the addition of a separate transmit transducer.[10] High-resolution multi-beam SONAR provides centimetre-detailed information about the seafloor. The traditional echo sounder can measure only the small portion of the seabed below the survey vessel. Modern acoustic techniques—called swath bathymetry—permit both precise water depth measurement and mapping of seafloor morphological features. Multi-beam system echo sounders, high frequency side-scan SONAR, sub-bottom profilers and Doppler SONAR can obtain extensive geo-morphological details of wide seabed portions and identify physical properties of sedimentary formations even below the sea bottom.

Another advanced active SONAR technology is the **Acoustic Doppler Current Profiler** (ADCP), which has applications for measuring speed and direction of ocean currents throughout the water column. ADCP technology is based on the Doppler effect.[11] The ADCP emits high frequency pulses that scatter off particles in the water. The frequency of the back-scatter echo is different depending on whether the particles are moving toward or away from the emitting source, and since the particles move at the same speed as the water in which they are in, the shift in frequency is proportional to the speed of the current. The ADCP has four acoustic transmitters that emit and receive pulses in four different directions. On large R/Vs one ADCP is permanently mounted on the bottom of the vessel's hull. The emitted pulses extend from the vessel down to the bottom of the ocean and an on-board computer processes the returned signals measuring the magnitude and direction of the current at different depths.

SCIENTIFIC AND COMMERCIAL APPLICATIONS

Following the previous distinction between (A) optic and (B) acoustic technology, the different applications of these two basic sensing technologies in the marine environment are examined separately.

Optic Techniques

Optic sensors—especially the ones operating at microwave length—have a number of oceanographic applications, including mapping sea ice, currents and surface winds, and detection of pollutants such as oil spills. There are many surface optical and bio-optical properties of interest to marine science that can be measured by optic remote sensor technology. Multi-spectral observations of these parameters permit to physically categorize water according to colours, giving colour pictures of the ocean (the so-called "ocean colour"). Ocean colour analysis is a method of assessing the health of the ocean by measuring its biological activity through optical means. The measurements of these parameters are assisting in better understanding the basic function and structure of the marine ecosystem. Even though biological sensors are less developed than physical sensors, optic technology has many applications in marine biology.

In plankton research, the most widely used bio-optical sensors are chlorophyll fluorometers and IR optical backscatters (Smart 2000). The LASER fluorescence technique used for determining the composition of *phytoplankton* populations is called flow cytometry. Flow cytometry identifies organisms by measuring backscatter signals and fluorescence of individual particles. It can detect and analyze even a single cell. Light-sheets, both single—such as the Optical Plankton Counter (OPC)—and multiple, are other techniques for obtaining information on size and abundance of plankton particles. UV fluorometers and other LASER sensors are widely employed to detect and measure pollutants at sea such as oil spills, produced water,[12] ballast water discharge, CO_2 and nutrients[13] (Griffiths 2005). UV fluorometers are able to determine the class of hydrocarbon and to map the dissolved oil distribution at sea. IR gas analyzers, which are either towed by ships or employed in moored buoys, can measure CO_2 concentration in the upper level of the water column. Fluorometers such as UV-absorption and IR-absorption spectrometers are employed to measure the state of nutrients in the ocean, the oil content of produced water and the chemical content of ballast water discharges.

The main parameters that are commonly measured at sea with optic techniques are:

1. The Diffuse Attenuation Coefficient (DAC), which is a descriptor of the penetration of sunlight in water and therefore a measure of water turbidity;
2. Suspended Particulate Matter (SPM). Since sediments in suspension transport all types of water pollutants and can even decrease the light penetration into the water, in turn affecting photosynthesis, water temperature and fish feeding, SPM is an important environmental indicator to determine primary production and pollutants at sea;
3. Coloured Dissolved Organic Matter (CDOM). Organic matter is a compound that forms from decomposing plant matter and has a major role in the underwater light climate. Organic matter absorbs light at the blue end of the spectrum, giving water a yellowish colour at high concentrations. Its optically active components, denoted as "yellow substance", are measured by absorption and fluorescence spectroscopy;
4. Chlorophyll-A is a parameter recorded to measure the concentration of *phytoplankton* pigments and other primary production indicators. It reveals *phytoplankton*'s abundance and water quality;

5. Macrophytes level. Macrophytes are aquatic plants, the presence or absence of which may indicate numerous phenomena such as water quality, nutrients concentration and, therefore, changes in fish population levels;
6. Nutrients level. Nutrients, such as nitrate, phosphate, silicate, nitrite and ammonium are of fundamental importance in biogeochemical studies. Some measures of nutrients, such as Dissolved Nitrate Concentration (DNC), can be detected with UV-absorption spectrometers. However, water sample collecting for chemical analysis—called "wet chemical"—is still the major tool for nutrients measurement in microbiological analysis;
7. Sea Surface Temperature (SST). The process of collecting SST data from radiometers is well established. The radiations absorbed by the surface and re-emitted as heat are detected and measured. This process is called thermal imaging. Data derived from TIROS, NOAA and the METEOSAT satellites produce SST charts with an accuracy of 0.5°–2° C in near real time;[14]
8. Sea state. Microwave sensors are able to measure with a high degree of accuracy parameters such as wave height, wavelength and direction, near surface wind speed and the microtopography of the ocean surface. Several ocean circulation features—both vertical and horizontal—including the location and evolution of frontal boundaries, up-welling areas, and currents and circulation patterns in general can be detected. Microwave techniques—particularly the use of RADAR altimeter—have applications regarding large-scale circulation features. The two main parameters surveyed are Sea Surface Height (SSH) and Sea Surface Roughness (SSR). SSH is determined through RADAR altimeter by transmitting high frequency signals to the Earth (over 1,700 pulses per second) and receiving back an echo from the sea surface. Changes in the SSH are used for predicting tides and storm surges. SSR is determined by collecting data such as surface patterns of currents, tides and waves;
9. Salinity. The microwave properties of the sea surface are functions of its physical and chemical state and the emissivity of seawater is related to salinity. Therefore, changes in salinity cause significant changes in the emissive brightness temperature of water. Hence, salinity can be determined remotely by measuring the emissive brightness temperature of seawater. However, the measurement of salinity from remote sensors is not operational yet;
10. Fish detection. In clear meteorological conditions, a LIDAR carried on aircraft flying at an approximate altitude of 1,700 m can detect fish at depths down to a maximum of 16 m. Another efficient optical technique used in fisheries research is Laser Line Scan (LLS). This is an electro-optic imaging technique producing underwater high-contrast light field imaging, the resolution of which approach that

of visual observation. Although the resolution of the images and the area covered vary according to water clarity, the details that can be observed with LLS are impossible to obtain with much lower resolution of acoustic techniques.

Acoustic Techniques

Numerous acoustic techniques are employed both in marine research and commercial ventures, providing powerful and non-invasive means for sampling considerable bodies of water in short periods of time. Acoustic sensors can extend observation from individual organism to entire populations of fish and plankton, covering areas ranging from few centimetres to hundreds of kilometres (ICES 2002). Acoustic methods are employed in fishery surveys to estimate fish stocks' biomass for management[15] purposes, in monitoring endangered ecosystems, in marine seismic surveys to discover oil and gas reservoirs, in mapping bottom topography, in marine archaeology, and in several other applications.

Multi-beam SONAR systems are employed to study the physical processes that shape the seafloor and therefore have applications in bathymetric measurements for digital mapping, in oil seismic surveys to detect reservoirs, and in marine archaeology to identify and describe shipwrecks. Hydrography, the science that deals with seafloor mapping, was revolutionized by the advent of such devices. Multi-beam bathymetric mapping systems are capable of mapping the seafloor down to several thousand meters depth and allow for a rapid and highly accurate coverage of large seabed areas. Onboard computer processing usually produces coloured 3D bathymetric maps with down to 5 m contour intervals.

The most advanced high-resolution profilers and algorithms for processing echo-acoustic data can be found in the field of seismic oil exploration. The full spectrum digital sub-bottom profiler for instance—which is a wideband FM profiler that generates cross sections of the seabed and collects reflection data over many frequency ranges—has greatly eased offshore oil and gas prospecting activity. ADCP, another recent application of acoustic technology, is used to measure the structure of the current. Many offshore operations require information on the structure of the current. Real-time data on current movements throughout the entire water column are usually required by the offshore oil and gas industry to monitor currents forcing water structures and to safely and timely plan underwater operations[16] at specific site-areas. In order to understand the ocean forces that will affect offshore structures, studies on current have been conducted since the early stages of the oil field development. Low-frequency ADCP at fixed points or towed by vessels assists in providing such information. Deepwater rigs are usually equipped with ADCP to monitor both underwater and surface currents.

Acoustic technology also has several applications in fisheries research. It is employed to obtain estimates of fish stock biomass and densities and in mapping the spatial distributions of fish. Echo sounders and SONAR have been in use as remote sensors for the last 50 years and are widely used also by fishing vessels because they are useful for the detection of fish and for stock estimation. Since the air-filled swimming bladder of fish is of different density than seawater, sound waves respond differently through fish than just water, and thus fish can be easily detected. Since each species of fish reacts differently to different sound frequencies, a multi-frequency echo sounder is able to identify different species of fish. The key detection parameter in these estimations is the fish backscattering cross section or Target Strength (TS). The backscattering TS is the echo strength of the returning signal and it responds differently for each fish species and size.

Many marine organisms—especially mammals—produce sounds that can be detected and recognized. Despite the major benefits that acoustic techniques provide to marine science, they are limited by platforms' constraints (Griffiths et al. 2001b). R/Vs typically have echo sounders mounted on the hull, in a drop keel or towed, therefore, the sensor lays several meters below the surface leaving the upper water column unsurveyed. Fish may tend to avoid the noise emitted from the engine of the R/V, thus many estimates of fish stocks may in fact be distorted. To overcome problems of fish avoidance and other behaviour responses to vessels' noise, more silent R/Vs have been built. However, usually the sensors are optimized according to the platform in which they are employed. The sensor's size, consumption of energy and weight are the major constraints in small platforms. Acoustic devices are usually heavier and of bigger size than optic sensors.

MARINE PLATFORM TECHNOLOGY

The vehicle—stationary or mobile—from which remote sensors operate is called a platform. Oceanographic research is usually conducted from a variety of platforms[17] such as R/Vs, stationary observation systems (underwater labs, moorings and buoys) and other mobile observation systems such as satellites, aircrafts,[18] balloons and submersibles. Although numerous state-of-the-art platforms are extensively used, R/Vs are still the main platform from which marine research is conducted. Modern R/Vs are designed as flexible multi-mission platforms that can be endowed with different instrument systems to suit specific project requirements. Large class R/Vs costs are usually based on 300-day-per-year use at approximately USD 20,000 per day (Committee on Exploration of the Seas 2003). Ocean exploration multi-purpose R/Vs are often equipped with a computer lab, swath mapping devices, SONAR, deep-towed camera and video systems, and TV-guided grab samplers. In addition, the R/V is the mother-platform from which other marine platforms, such as submersibles and buoys, are usually deployed. The cost for secondary platforms is of course not included in the estimated cost of the bare R/V. Significantly increased costs are associated with the employment of underwater vehicles, especially those with deep-water capabilities. The amount of spare parts and test and repair equipments to be carried onboard is extensive and can therefore add significant costs to submersible employment.

Mission targets and consequent operating-depth requirements determine the packaging of instrumentation and components—such as buoyancy, electronics housing and other pressure-resistant components, specialized tooling, video, navigation, and sensor and sampling equipment—required both for the underwater vehicle and the support vessel. Regarding tethered underwater vehicles, mission requirements in turn determine the number and size of power conductors and fiber optic cables, which affect the characteristics of the umbilical, such as its complexity, overall diameter, length and weight.

The data transfer from the vehicle to the vessel is achieved via a fiber optic cable, while the energy is transmitted via a co-axial cable. Regarding the support vessel, the type of underwater vehicle employed determines the type of control centre and of the surface handling system, including the winch used to store, launch, and recover the underwater vehicle.

One limit of R/Vs is the physical region in which the survey can be conducted. For instance, under-ice areas present sampling difficulties, since equipped R/Vs can penetrate ice but disrupt the environment of interest at the same time. Due to the physical attenuation of sound—particularly at higher frequencies—in the marine environment, deep-sea areas are also difficult to survey in details from R/Vs. For these reasons marine research extensively employs remote platforms for scientific measurements, especially submersible vehicles that can collect data from sites that would otherwise be impenetrable.

Underwater Vehicles

Underwater vehicles are of fundamental importance for directly accessing the marine environment. In the mid-1950s, scientists began to discuss the potential of deep-sea research submersibles and, in the early 1960s, the first deep submergence research vehicle was built in the United States. Since the mid-1970s, manned and unmanned submersible vehicles have increasingly and helpfully supported marine research programs. As a result, there is a variety of submersibles: Human Occupied Vehicles (HOV), Remotely Operated Vehicles (ROV), Autonomous Underwater Vehicles (AUV) and deep-diving Hybrid Remotely-Operated Vehicles (HROV), which can operate either in ROV mode or AUV mode. The main difference when operating in ROV mode is the fiber optic tether, which is quite smaller than the traditional cable system. HROV autonomously collect and transmit data to the mother station onboard the R/V. When an area of interest is identified, the vehicle can be switched to ROV mode to look more carefully at the details.

The cost of underwater vehicles and other marine technological equipment has a trend that is similar to other high-tech devices, such as microelectronics, robotics, genetic engineering and biotechnology. Products are very expensive at the beginning because the implementation of a prototype involves a long and costly research program, however, once the production is well established costs begin to decrease. Submersible costs are in large part related to their depth capability. As a result of the extreme pressure and the expensive equipment needed to handle it, costs for submersibles' gears increase nearly logarithmically below 6,500 m. However, around 97 percent of the ocean can be accessed by existing undersea vehicles, which have a maximum depth capability of about 6,000 m. Underwater vehicle manipulators are also of great importance. They range from simple joystick-controlled grabbers to more costly, complex, and sophisticated systems donned of multiple tool options that use master-slave control functions to have superior handiness. It is worth noting that insurance companies do not insure many oceanographic gears.

Human Occupied Vehicles

Human occupied submersibles have five main components and systems: the pressure hull (which can have different shapes); the view port (usually of spherical or conical shape); the communication system; the power system; and the propulsion system. Unlike other underwater vehicles, HOVs have the additional requirements for life support and consequently they must display a complex safety system. This, in turn, requires the forging and welding of a 7–8 cm-thick titanium personnel sphere, which is particularly complex and costly to obtain. However, getting a structure strong enough to resist the high pressure at depth is not the only problem, another critical element in submersible construction is the balance between strength and weight. HOV weight can range from 5 to 25 tons. Apart from titanium, other materials used to build the pressure hull are ceramics, and graphite-based metal alloys and acrylic, the latter has a remarkable potential for future use.

The logarithmic scale of increasing pressure underwater makes safety requirements increase accordingly and so do costs. As a result, a HOV for shallow-water research costs between USD 200,000 and 300,000, while a HOV to work at depths of 6,000 m cost more than USD 20 million (Committee on Exploration of the Seas 2003). Russian submersibles *Mir I* and *II*, constructed some years ago by a Finnish company, are capable of diving down to 6,000 m and had a cost of about USD 25 million each. The new US HOV *Alvin*,[19] able to dive to depths of 6,500 m, when completed in 2008, will be the most capable deep-sea research vehicle in the world. Its four-year design and construction project is expected to cost more than USD 20 million. A full-ocean-depth HOV capable of reaching the 3 percent of ocean floor below 6,500 m would cost three to five times more (see Table 2.1).

The first deep submergence research vehicle—*Alvin*—was placed in service at the Woods Hole Oceanographic Institution (WHOI) in 1964. It could reach depths down to 1,900 m with dives lasting up to a maximum of 10 hours. Since that first achievement, underwater vehicles technology has advanced tremendously. Apart from the depth the vehicles can reach, all the sensors, observational and manipulative apparatus have extremely improved. Today sophisticated vehicles are equipped with precise measurement sensors that routinely acquire optic, acoustic, and chemical data. Cameras and video camera equipment provide images of the underwater world and robotic arms manipulate the surrounding environment. Most underwater research involves the use of deep diving submersibles and various deep-towed instrument packages and, consequently, there is a wide range of underwater vehicles available for marine research. Presently there are about 40 commercial and scientific manned submersibles operating at sea in the world.[20] However, only five deep-diving HOVs are available. Submersibles commonly provide accommodation for two pilots and one scientist and are launched from a mother vessel. The launch and recovery system itself can cost more than USD 300,000. Each dive usually lasts between 8 and 10 hours, however, dives up to more than 16 hours have been reported.

Even though manned submersibles have the advantage of direct human observation and manipulation, they have several disadvantages compared to unmanned underwater vehicles (UUV), such as shorter dive time, high costs, and risk for both pilots and scientists. For these reasons researchers are focusing on UUVs development. The advancement of UUV technology—such as highly controllable manipulators, screen at a control station, side-scan sonar for object location and obstacle avoidance, sub-bottom profiler etc—has proven to be valuable to both the civilian and the military sectors. The needs of the marine commercial industry, including the offshore oil and gas sector, the search and recovery sector and the cable sector, are mainly related to the following activities: pipeline and cable surveying (both inspection and routing); offshore platform IRM (Inspection, Repair and Maintenance); wellhead and satellite wellhead IRM; seabed object detection and classification; sub-sea construction; and wet salvage. For the requirements of many of these underwater operations, the recent development in UUV payloads—coupled with the considerable difference in cost—provide now little reason to employ manned submersibles.

Table 2.1 Human Occupied Vehicles for Scientific Research and Exploration

HOV	*Operator*	*Maximum Operating Depth (meters)*
Shinkai 6500	JAMSTEC, Japan	6,500
MIR I and II	P.P. Shirshov Inst. Oceanography, Russia	6,000
Nautile	IFREMER, France	6,000
Alvin	NDSF, WHOI, USA	4,500
Cyana	IFREMER, France	3,000
Pisces IV	HURL, USA	2,170
Pisces V	HURL, USA	2,090
Shinkai 2000	JAMSTEC, Japan	2,000
Johnson-Sea-Link I and II	HURL, USA	1,000
Deep Rover 1002	James Cameron	1,000
Deep Rover	Nuytco Research Ltd, Canada	900
Remora 2000	Comex, France	610
DeepWorker 2000	Deep Ocean Expeditions	600
Jago	Max Planct Inst., Germany	400
Delta	Delta Oceanographics, USA	370
Clelia	HBOI, USA	300
Thetis	Greek National Centre of Marine Research	300

Note: JAMSTEC (Japan Marine Science and Technology Centre); IFREMER (French Research Institute for Exploitation of the Sea); NDSF (National Deep Submerged Facility); WHOI (Woods Hole Oceanographic Institution); HURL (Hawaii Undersea Research Lab); HBOI (Harbour Branch Oceanographic Institution).

Source: Committee on Exploration of the Seas (2003). Reprinted with permission from the National Academies Press, Copyright 2003, National Academy of Sciences.

Remotely Operated Vehicles

Tele-operated systems with components such as manipulators and mobile devices are in general developed and used for extending human ability to perceive and interact in an environment that is too distant, dangerous, costly or otherwise inaccessible (Habib 2000). Remotely tele-operated underwater vehicles represent a typical example of systems that operate to extend the human's sensory perception and manipulative abilities into the harsh and hazardous marine environment. Thanks to payload systems that combine acoustic sensors with robotic arms, video, and other advanced imaging devices, ROVs can make detailed multidisciplinary surveys of large areas of the seabed and in the water column, giving marine scientists the opportunity to study a broad-scale distribution of physical and biological features. While maintaining navigational precision on the order of centimetres, ROVs can manoeuvre and sample at big depths, allowing comprehensive observation at an excellent scale of resolution and multimodal data collection.

ROVs are used to perform tasks that are too expensive, unpractical or not feasible with other platforms employed in marine field methods, such as HOVs, surface acoustic and satellite remote sensing. ROVs have been used to discover the wreck of the HMS *Titanic*,[21] to explore hydrothermal vent fields, to conduct mid-water chemical, biological and physical oceanography experiments and to perform under-ice surveys in the Arctic and Antarctic Oceans (Bachmayer et al. 1998).

The past 20 years have seen a blossoming of ROV technology, with many applications for both offshore oil and gas industry (in support of survey, drilling and production operations) and deep-sea research. Currently about 90 companies and institutions worldwide are developing underwater ROVs, providing the market with more than 300 different ROV types, establishing a fleet of about 3,000 units (Clauss and Hoog 2002).

ROVs are tethered to the mother ship by an umbilical cable. The development of the fiber optic system has greatly improved the solution of cable-related problems. Fiber optic cables employed in ROVs have a diameter of about 2 cm and a length that varies from about 200 m to 10 km. The umbilical provides both power for and control of the vehicle and continuously transmits data to the control station onboard the vessel. ROVs are launched from a crane or travel lift onboard the mother ship. The operator controls the ROV from the mother vessel using joysticks, video monitors, and several instrument readouts. Commercially available ROVs range from small, portable units used for shallow-water inspection to the heavy, work-class, deep-water ROVs used by the offshore oil and gas industry and the military. Weight ranges from a minimum of about 1 ton to 9 tons. Small ROVs are usually powered by electric motor thrusters of less than 20 Hp and operate in depths of less than 300 m. This kind of ROV is relatively inexpensive, prices ranging from USD 10,000 to USD 100,000 (Committee on Exploration of the Seas 2003). ROVs are used in a variety of marine scientific field researches, for underwater facility inspections, marine archaeology, sub-marine cable placement, and repair and other observational tasks. Medium-class ROVs cost around a million dollars and operate at depths of 1,000–2,000 m. They weigh a metric ton or more and require additional equipment to operate, such as an overboard-handling system, winches, heavy generators and a sophisticated control system. Therefore, they are not readily portable but are typically semi-permanent installations on support vessels. On the other hand, they can carry a variety of payloads,[22] including acoustic sensors such as SONAR or echo sounders, non-acoustic sensors such as magnetometers, sampling equipment, manipulators, and special devices such as water jets and cutting tools, or even military applications such as weapons and countermeasures for mine removing.

A typical medium-class ROV is equipped with electro-hydraulic arms and with video eyes. Depending on the specific characteristics and needs of each single mission, ROVs can also be kitted out with biological and physical sensors and with a set of cameras and video cameras.

Marine researchers use these vehicles for video exploration and photographic document support, instrument placement, and oceanographic data recording, as well as sample gathering. Heavy-class ROVs provide the maximum underwater performance. They are equipped of up to 500 Hp and can reach depths of about 5,000 m. They can carry heavy payloads and a wide variety of additional tools. They cost up to USD 2.5 million and, consequently, they are seldom used outside the offshore oil and gas sector and the military (Committee on Exploration of the Seas 2003). The price of ROVs with very deep capability is much higher. The *Jason II* ROV, projected by the Deep Submergence Laboratory Monitoring of Massachusetts,[23] costs an estimated USD 4 million to construct, while the Japanese full-ocean-depth ROV *Kaiko*, lost at sea during a typhoon in the Pacific Ocean in March 2003, had a cost of more than USD 60 million.

Autonomous Underwater Vehicles

AUVs are self-contained, usually torpedo-shaped, unmanned and unleashed underwater robotic vehicles able to travel several thousand kilometres at a time. They are no longer the fantasy devises of science fiction writers or engineering curiosities without any practical application. AUVs have been under development since the 1970s and play an increasing role in underwater research and exploration. This technology consents a significant cost saving in the budget of the marine stakeholders involved in underwater

Table 2.2 Remotely Operated Vehicles (ROVs) for Scientific Research and Exploration

ROV	*Operator*	*Maximum Operating Depth (meters)*
Kaiko	JAMSTEC, Japan	11,000
Ropos	CSSF, Canada	6,000
Jason	WHOI, USA	6,000
Victor	IFREMER, France	6,000
Dolphin 3K	JAMSTEC, Japan	3,300
Robin	IFREMER, France	3,000

Source: ISA (2002), reprinted by permission. All rights reserved.

operations. Therefore, over the past decade, more than 170 AUV prototypes have been developed around the world, testifying the present interest in these vehicles. Consequently, recent years have seen a significant increase in the use of AUVs in operational missions, ranging from scientific (marine research and ocean mapping) to commercial (pipeline inspection and repair, offshore oil and gas underwater operations, sub-sea mineral explorations) and military tasks (mine counter measures and surveillance).

Regarding marine research, an AUV employed in operational oceanography is a remote, versatile and adaptable data-gathering platform donned with sensors. However, AUVs should not be thought of as just an alternative to traditional data-gathering methods such as R/Vs, drifting buoys or moorings, since with the appropriate payload AUVs can obtain measurements of what would either be very difficult, time consuming or impossible (Griffiths et al. 2001b).

As AUVs are set free at sea, the primary requirement is to ensure the survival of the vehicle. High technical reliability and ability to cope with unforeseen circumstances are therefore a must in AUV design. Dangers such as collision, major pressure vessel leak, and battery short circuit can easily lead to the loss of the vehicle. However, losing an AUV in the open ocean is an uncommon event because most vehicles have visual, radio, SONAR and satellite tracking systems that enable operators to find the vehicle if it is lost on the surface or trapped underwater. In addition, for the vast majority of faults the emergency abort system is able to sense problems and to take appropriate counteractions automatically. Regarding the reliability of these vehicles, researchers estimated within a panel of 118 missions covering 2,519 km that the probability for an AUV high impact fault when underway is about 0.07 percent (Griffiths et al. 2001a).

The structure of AUVs is made up of the hull and a ballast system to maintain the vehicle neutrally buoyant. The hull contains all the other components:

1. Propulsion system;
2. Control or navigation system;
3. Payload.

The propulsion system is usually made up of conventional screw-type propellers, while an energy storage high-density battery system provides the essential power supply. However, since low noise, vibration, and high stability are required for several marine measurements, jet-propulsion systems are beginning to be extensively used because of their lower noise and subsequent lower disturbance. Usually about 25 percent of the vehicle's hull is reserved for the power source, even though this may vary according to the operational range and type of mission for which AUVs are used. Electrochemical power sources for AUVs may be divided into four different groups (Størkersen and Hasvold 2004):

1. Standard rechargeable batteries based on different metals, including conventional lead and nickel batteries, silver-zinc and lithium batteries. Alternatively, more expensive high-density battery systems such as manganese alkaline or lithium-based cells[24] can be employed. These types of batteries are located inside a pressure container and work at normal pressure;
2. Pressure compensated batteries such as lithium-polymer ambient pressure batteries and aluminium-hydrogen peroxide batteries (semi fuel cells). These types of batteries work at ambient pressure, but are electrically insulated from the seawater;
3. Seawater batteries, which make use of the oxygen contained in the seawater to generate energy;
4. Fuel cells batteries, operating on hydrogen and oxygen fuel.

Autonomous vehicles should ideally be able to operate continuously over many days (or even weeks) and at the same time the overall size of the vehicle body should be limited. Therefore, to exploit the true operational benefits of AUVs one of the technological areas that needs to be developed in this field is high-energy density power sources. Endurance requirements are easier to meet in large vehicles than in smaller ones because of the volume to store the power source. Two technologies for pressure-tolerant battery packs are under development: individual lithium polymers cells; and semi-fuel cells. They are insensitive to water depth and thus especially attractive for deep water AUVs. Lithium polymer cells batteries have a performance similar to the conventional lithium-ion batteries; however, they have the advantage of operating at ambient pressure. Regarding hydrogen-oxygen battery packs, although fuel cells technology[25] is approaching maturity in some applications, it is only in its early stages of development in AUV propulsion systems. The extension of this technology to AUVs has great potential for the future, because its main feature is the reduced size of fuel cells. To satisfy endurance requirements, an alternative approach to get more powerful energy storage is simply to refuel the vehicle during its missions. However, this is not as simple as it looks, since especially equipped offshore docking stations would have to enable the AUV to moor, upload and download data, and recharge the batteries.

The control system that steers the vehicle is made up of hydrodynamic fins, which rely on forward motion to generate the forces required to change direction. To better control their motion—especially at low speeds—many AUVs employ joint fins with through-body thrusters. Control problems present several difficulties, such as presence of external forces of disturbance (waves, currents) and natural and artificial obstacles. Therefore, it is necessary to provide the AUV with the ability to first track its own position and then fix the obstacle position by using its relative distance and orientation with respect to the AUV.

The components required for operations such as obstacle avoidance at surface are usually SONAR, computer systems and a long-range navigation and positioning system—typically, a satellite Global Positioning System (GPS).[26] However, since GPS/DGPS (differential GPS) signals do not work in an underwater environment, standard acoustic navigation systems are used when the AUV dives down the surface. A development alternative to GPS is Simultaneous Localization And Mapping (SLAM), which allows the AUV to map the environment and then use the map to localize itself. Current state-of-the-art algorithms employed in obstacle avoidance systems can provide the highly accurate paths required for safe navigation using artificial intelligence incremental search techniques. The success of these methods depends on the AUV's understanding the representation of the surrounding environment. This, in turn, depends on the certainty of its position when obstacles are observed and mapped (Patron and Tena Ruiz 2006).

Prompt real-time reactions are especially required to avoid collisions when the AUV navigates inside dynamic environments. Several control devices have been developed for controlling AUVs' motion and minimize navigation errors. Among them are Doppler Velocity Log (DVL), aided Inertial Navigation Systems (INS), supervisory control, neural network control, self-turning control, sliding mode control and fuzzy logic control. These last two strategy controls have recently been joined in order to obtain a more efficient control system (Guo et al. 2003).

Another developing technology in AUVs' motion control is based on biomimetics—i.e. the use of biologically-inspired technologies. Concerning AUVs, this means to replicate the locomotion of aquatic animals. In Japan for instance, the use of a mechanical pectoral fin as a device for manoeuvring and stabilizing an AUV is underway (Kato and Suzuki, 2006) as an alternative way to the typical AUV motion controlled by thrusters and wings. In this system three servomotors are used to generate three different types of motion: a rowing motion; a feathering motion; and a flapping motion.

The payload system contains the instrumentation apparatus to trace the pre-set parameters of interest to be detected. According to its dimension, an AUV can be configured with different types of optic and acoustic sensors, such as a radiometer, echo sounder, multi-beam SONAR, ADCP, fluorometer etc. Still or videocameras may also be part of the payload to transmit computer-aided 3D imaging vision to the mother vessel.

AUVs employed in marine research are optimized to survey oceanographic variables of interest over wide areas and can be set loose for hours to months at a time, scanning the seabed or the water column along preprogrammed paths and then recovered by the support ship for data downloading. Alternatively, they can be set to periodically pop up to beam their data to satellites in order to be re-transmitted to ground stations. Large class AUVs employed in bathymetric surveys[27] and pipeline inspections join long mission ranges of several thousand of miles with dives deep down to 600 m. They weight up to 1,300 kg, have a length of about 5 m or more and cost more than USD 200,000 (Committee on Exploration of the Seas 2003). For instance the construction cost of the AUV *Hugin 3000* is about USD 3.5 million with expected additional costs of USD 2 million to fully outfit the system with payloads and about USD 3 million a year for its maintenance.

In order to minimize or avoid expensive ship-time and to collect data offshore, the use of small portable AUVs is becoming routine in scientific research, sometimes even deploying them as a stand-alone platform from the shore or from a rubber dinghy. Small AUVs require less formidable support teams and are far easier to deploy and recover, but they have a short range of action. However, even if AUVs are launched and recovered by a support vessel, there are significant cost savings, since the ship does not need to be in attendance but is free to pursue other scientific objectives while the AUV is making its survey.

AUVs have another advantage over ROVs as the latter cannot cover as much distance as AUVs in the same amount of time (AUVs can run at three knots) because of the ROV tether, which can be even thousands of meters long and three centimetres thick. The tether produces drag on the vehicle, making it less manoeuvrable and vulnerable to entanglement, especially in difficult areas. Furthermore, to operate a towed system in deep water requires big, expensive level winding winches, expensive cables and a dynamically positioned vessel that can double the costs per day, while an AUV just needs a crane to swing it out and away from the hull of the mother ship and drop it into the water.

Satellite Platforms

Satellites with remote-sensing capabilities are another important and helpful platform supporting marine research. Observation from space, initially developed for meteorological purposes, has revolutionized oceanographic observation. Since 1957, when the first *Sputnik* was launched by the USSR, thousands of satellites have been launched from numerous countries. From the first images with poor spectral and spatial resolution of the late 1960s, ocean-colour satellite imaging today gives an unprecedented global overview of the variability of biological production in the oceans. Some satellites—called geostationary—orbit in synch with the spin of the Earth and therefore are fixed in a set position above the planet's surface. Others orbit along different tracks and can take global snapshots as they pass over different regions of the Earth. Satellite remote sensing gives marine scientists the possibility to detect different aspects of ocean conditions such as water temperature, chlorophyll levels and oceanic waves and currents. Other applications of satellite observation include weather reports, search and rescue operations,[28] bathymetric measurements and environmental monitoring. Remote sensing techniques from satellite platforms are also employed to track animals tagged with beacons that transmit information to the satellite. From the satellite the data are then transmitted to a ground station.

The 1978 Nimbus-7 satellite was the first orbiting platform to carry a payload of sensors exclusively planned for monitoring the oceans: the Coastal Zone Colour Scanner (CZCS) system. The CZCS sensor had six spectral bands in the visible, near-IR and thermal portions of the electromagnetic spectrum. Its primary objective was to observe ocean colour and temperature. MOS-1 and MOS-1b, two Marine Observation Satellites launched by Japan in 1987 and 1990, carry three different types of sensors: a four-channel Multi-spectral Electronic Self-Scanning Radiometer (MESSR); a four-channel Visible and Thermal Infrared Radiometer (VTIR); and a two-channel Microwave Scanning Radiometer (MSR). The Sea Star spacecraft was another satellite specifically designed to monitor ocean phenomena such as ocean influences on climate processes (the cycles of carbon, sulphur and nitrogen) and ocean primary production (*phytoplankton*). The SeaWiFS (Sea-viewing Wide-Field-of-View Sensor) is the advanced sensor system for ocean monitoring onboard the Sea Star. It consists of eight spectral bands of different wavelengths.

Satellite observation is performed with optical devices and has therefore a vertical limit of penetration in the water; only the surface and the upper level of the water column can be detected. Consequently, satellite monitoring can presently detect only one biological parameter at sea—i.e. biological productivity, detected through *phytoplankton* biomass pigment analysis. The electromagnetic radiations of chlorophyll pigments give information on the *phytoplankton*'s density in surface waters. This information permits in turn to quantify *phytoplankton*'s concentration and variability and thereby to assess its productivity. These advances allow scientists to have a global picture of the condition of *phytoplankton*.

Before satellite observation was employed, only *in situ* measurements of bulk chlorophyll concentration were possible. They were made first with the fluorometers and, more recently, with flow cytometers. Even if 2D observation from space can produce systematic global ocean data of the horizontal parameters to be detected, the use of infrared sensors—the technology employed by Landsat satellites to measure SST—is limited by cloud covering in the atmosphere, which is especially frequent at high latitudes. However, microwave sensors, which started to operate in the late 1970s, are independent from meteorological variability.

State-of-the-art satellites carry on board multi-frequency scanning microwave radiometers, which allow extensive data gathering of geophysical parameters such as SST, SSR, sea surface wind speed etc (Bhatt et al. 2004). For instance Topex/Poseidon, a joint French-US satellite mission to monitor global ocean circulation patterns, can measure the height of the ocean's waves with an accuracy of a few centimetres. Topex/Poseidon orbits 1,336 km above the surface of the Earth and every 10 days makes a map of the height of the waves. Scatterometer RADAR (used to measure sea winds) can detect even finer details and is used to locate submerged banks and reefs by measuring the swell of the waves above them. Bathymetric surveys to map shallow sea areas (maximum depth of about 30 m) are now possible using satellite microwave imagery. Data from satellites are frequently compared with information gathered from buoys located at sea to confirm the reliability of the data.

Apart from navigation, satellite RADAR altimetry finds application in other marine activities. Regarding fishery research, it is known that fish often concentrate along oceanic fronts and eddies to feed. Since these fronts and eddies can be detected from altimeter-derived data, satellite altimetry can be employed to discover where these fronts are located and monitor fishing stocks for fisheries management purposes. Satellite altimetry also has applications in other numerous offshore commercial sectors because marine activities are notably sensitive to weather condition at sea. Apart from general navigation purposes, cable installing companies and offshore oil and gas companies particularly rely on weather information for the safety of their operations. Other oceanographic applications of satellite observation include environmental monitoring, such as identifying oil spills and locating algal blooms, and sea ice mapping.

Other Marine Platforms

In addition to underwater vehicles and satellites, a whole range of marine platforms are making scientific research and marine commercial industries more efficient and cost-competitive. Other marine platforms include ocean observatories, underwater laboratories, drifters and gliders, moored profilers and benthic landers.

Ocean observatories are underwater facilities able to ensure a permanent presence on the seafloor in order to collect multidisciplinary physical, chemical and biological data. Capable of performing autonomous, long-term oceanographic observations, these platforms give scientists the opportunity to monitor both sporadic short-term events and long-term trends. Several academic institutions in Europe, Japan and North America are planning to establish their own permanent ocean observatories. Long-term deep-sea observatory stations have been located at water column and seabed study sites to allow observations of biota and *in situ* experiments.

GEOSTAR (Geophysical and Oceanographic Station for Abyssal Research) is a joint European project concerning a benthic (seabed) station. It consists of two main components: the deep-sea observatory, and the intervention tool, which is a ROV named MODUS used to deploy, recover and service the station (Clauss and Hoog 2002). Another interesting project is NEPTUNE,[29] which has the objective of coordinating many observatories in a network using submarine fibre optic cables. Since communication occurs between numerous seafloor instruments and stations and land sites, the submarine infrastructure required to power and supply data communication network includes thousands of kilometres of medium voltage fibre optic cable and terminations, transmission equipment, communication interfaces and other TLC instruments (Waterworth and Chave 2004). However, since cable connections are subject to damage by trawling and anchoring, acoustic wireless modem techniques can be usefully employed to insure data transmission without risk.

More than 65 **underwater laboratories** placed around the world by governments and scientific organizations are being used for sustained observations and measurements. The environment inside the underwater lab is pressurized in order to equalize the pressure of the surrounding ocean. Seabed observatories are linked by cable and moorings and are donned with satellite communication system (Census of Marine Life 2001). A generator located in a life-support buoy, which is moored at surface and tethered to the lab structure, provides electrical power. Aquanauts spend up to 60 days living and researching in the seabed in saturation conditions. There are also a few underwater stations located on the deep seafloor, such as the Japanese Real Time Deep Sea Floor Observatory off Hatsushima Island in Sagami Bay.[30] In these observatories, real-time extended observation methods are adopted to gather time-series. Such methods are currently being developed in Canada, the USA (H2O in Central Pacific, MARS in NE Pacific, Monterrey Bay) and Europe (ESONET and ANIMATE in the North Atlantic).

Drifters and gliders are small portable platforms (buoys) that can carry a limited payload of sensors. However, the system is scalable, since the small structure of these autonomous platforms permits the employment of as many units as necessary to cover the physical dimension of the area to be surveyed. They are much cheaper than AUVs and, as a result, are often employed in fleets. They are usually programmed to surface periodically in order to interface data with a satellite. **Drifters** (or drifting buoys, a.k.a. floats) are small autonomous platforms that provide researchers with information about ocean circulation patterns in real time. Drifters' design is quite simple.[31] Shallow water drifters weight only about 10 kg and have five major components: The body, the sails, the floats, the sensor payload and the transmitter package. The body is usually a waterproof tube approximately 1 metre long and 30 cm wide that contains the sensors array and the data transmitter. Extending out from the sides of the tube are four pairs of arms, each of which has approximately 1 square meter sail stretched between them. These underwater sails act in the same way that sails on boats do, however, rather than catching the wind, they catch the water, moving the drifter along with the prevailing currents. In order not to be blown along by the winds, the drifter is suspended several meters beneath the surface by ropes attached to four small floats. Drifters can be deployed from a ship or from an airplane and—apart from sensors to measure ocean circulation patterns—they may also house sensors that measure other ocean properties, such as surface temperature, wind, ocean colour, pressure, or salinity. The transmitter sends data to a satellite that calculates its position and relays the information to a receiving station.

Underwater **gliders** (or profiling float) are basically drifters with wings. The wings are necessary to provide lift and allow the glider to move also horizontally. Gliders can sample the ocean both vertically and horizontally. Power consumption of underwater gliders is minimal because of the design of their propulsion system. Electric energy is required only to activate a pump that changes the buoyancy of the glider. Once the required buoyancy is established, the wings convert changes in buoyancy into a low forward motion. Gliders can therefore achieve long-range missions with a small power source. Gliders are especially suited for deep-water and long-lasting measurements. Three gliders are operational for marine research and are commercially available: Spray; Slocum;[32] and Seaglider. They appear as cylindrical aluminium cases ranging in length from 1.5 m to 2 m and have a diameter between 20 cm and 30 cm. They displace around 50 kg and can carry around 5 kg of payload. Operating depth ranges between 200 m and 1,500 m, while cost ranges from USD 50,000 to USD 70,000 (Rudnick et al. 2004).

Moored profilers are semi-fixed platforms that crawl the water column up and down flawing along a moored guide wire while recording the parameters that their sensor payload detects. **Benthic landers** are self-contained platforms that can work remotely on the seabed, often thousands of meters below the surface. Landers are deployed with attached iron ballast. Once on the seabed, the required experiments are carried out remotely, controlled by an on-board computer. To recover the lander, an acoustic command is sent from the surface to release the ballast, allowing the lander to float back to the surface.

Other technological equipments in use today in the marine environment include: DGPS navigation system connected to underwater acoustic transponders, employed to keep a vessel still at a fixed point in the ocean; underwater communication systems; communication systems linking satellites with other marine platforms; marine leach mining systems and free fall grab (FFG) for sampling; fast stable ship technologies; underwater jet pump transfer systems; and new high tech metal and ceramic alloys. These technologies, which reduce costs, thus increasing the competitive advantage of offshore producers, are being incorporated into every aspect of offshore oil and gas and mineral exploration (Wiltshire 2001). They are expected to soon become in use in scientific research and in other marine commercial sectors. Other examples of sophisticated monitoring instrument packages include time-lapse cameras and temperature arrays, seismic-monitoring equipments, ground-deformation monitors, fluid-pressure monitors and chemical scanners.

Another recently developed technology is the Autotracker system, which is a fully autonomous pipeline tracking and surveying payload that can be integrated in any kind of underwater vehicle. It can be used to inspect more efficiently underwater pipelines, since its acoustic sensors—side scan and multi-beam SONAR—are capable of autonomously detecting a pipeline's location and following the pipe. Compared with traditional methods, the predicted saving for pipeline inspection is expected to be up to 30 percent (Patron et al. 2006). Another powerful tool for scientific research is the Unmanned Surface Vehicle (USV). This kind of marine vehicle is suitable for a wide range of tasks such as seabed surveying, carriage and deployment of daughter vehicles, tracking of underwater objects and can even be used as a platform for carrying sensors (Hook 2006). USVs have been developed all around the world in a vast range of hull types and applications. They range in size from about 1 m to 20 m in length, weighing from about 50 kg to 30 tons and having a speed of 5-50 kts. USVs' control system can either be through direct radio control of the vehicle by an operator with visual feedback or an autonomous control system.

3 Offshore Oil and Gas

GENESIS AND FIGURES

Natural petroleum is a hydrogen-carbon compound made of a mixture of highly reduced hydrocarbon molecules mostly found in the liquid phase.[1] Hydrocarbons are traditionally believed to have a spontaneous genesis: biological detritus are transformed into hydrocarbon molecules through a process of decomposition in condition of high pressure.

This theory states that long time ago dead organisms sank into the sand and mud at the bottom of the ancient seas. Over a process lasting millions of years, organisms decayed forming carbon-rich compounds that accumulated in sedimentary layers where there was little or no oxygen present. These organic layers mixed with sediments, forming fine-grained shale, or source rock. As new sedimentary layers deposited, they exerted intense pressure and heat on the source rock. Heat and pressure distilled the organic material into crude oil and natural gas, which flowed from the source rock and accumulated in thicker and more porous limestone or sandstone rocks called reservoirs. Movements in the Earth trapped oil and natural gas in reservoirs of impermeable rock such as granite or marble.

During the past 50 years, Russian-Ukrainian scientists developed an alternative a-biotic theory of hydrocarbon genesis. Nikolai Kudryavtsev first enunciated the so-called "modern petroleum science" in 1951. Having analyzed the spontaneous genesis of hydrocarbons in the framework of the chemical thermodynamic-stability theory, scientists found out—given temperature and pressure at Earth's near-surface crust—the effective prohibition of transformation of biological molecules into hydrocarbons heavier than methane. The hypothetical pressure needed to transform biological molecules into hydrocarbon is 30 kbar, which correspond to a depth of 100 km, which of course is not the case. Consequently, oil and gas formation is set in a wider context of geological processes, including not only the sedimentary section, but also the deep interior of oil and gas-bearing regions located in the Earth's crust and mantle (Kenney 1996). Put simply, petroleum is a primordial material erupted from the Earth's great depths.

Apart from their controversial genesis, since the last century oil and gas have become of fundamental importance for the world economy. The oil and gas industry currently satisfies 63 percent of the world's primary needs for heating, lighting and transportation (WSSD 2002), not to mention that more than 500,000 industrial and consumption products are derived from petroleum.[2] Translated into total consumption, this means that in the year 2000 75 million barrels[3] (mb) of oil were demanded everyday, together with 1.75 billion cubic meters of gas. In 2005 total world demand for oil rose to 83.6 mb per day, nearly 30 billion barrels (bb) of oil per year (OCSE 2006).

During the last decade, the rate of growth of oil and gas consumption has been about 2 percent per year. Wells can produce oil, oil and natural gas, or natural gas only. Natural gas production is closely related to crude oil, since almost all oil deposits contain also natural gas. Therefore, natural gas is often a by-product of oil because gas comes out of the solution as it endures pressure reduction going from the reservoir to the surface. However, natural gas is often found without oil. Since hydrocarbons are basic but non-renewable resources, it is of utmost importance to determine the quantity available for future consumption.

In estimating oil resources different concepts are used. The geologic concept, called "original oil-in-place," measures the total oil resource base found in nature without regard to its recoverability. The economic concept refers instead to the proven oil reserves, called "recoverable conventional oil resource" [4] and indicates what percentage of the total oil resource can be recovered given the current economic conditions and the technological capability. Recoverable conventional oil resource includes four items:

1. Cumulative production;
2. Proved or remaining reserves—i.e. oil that has been discovered, not produced but has a 90 percent confidence of recoverability. 77 percent of world's remaining oil reserves belong to government-owned companies;
3. Undiscovered conventional reserves—i.e. oil assumed to be in place by geological knowledge and to be found through exploration;
4. Reserves growth in discovered fields—i.e. increase in reserves resulting from: (A) technological factors that enhance the field's recovery rate; and (B) discoveries of new reservoirs within a discovered field.

Therefore, estimates of the world recoverable conventional oil resource vary according to the assumptions on recoverability and to new discoveries. Estimates have always trended upward over the years. The constant increase in estimates' figures is partly due to technological advances in the production process, which allow the economical viability of more and more reservoirs. Starting from the 600 bb estimated in the early 1940s, there has been recently a blossoming of estimates from different institutions and professionals of the sector. In 2004 according to OPEC there were 1,147 bb of proven crude oil reserves, of which 78.4 percent was in OPEC member countries (OPEC 2005). For the year 2000 the magazine *World Oil*[5] gives a figure of 1.1 trillion barrels (tb); British Petroleum[6] of 1.2 tb; *Oil and Gas Journal*[7] of 1.3 tb; IHS Energy[8] between 1.3 and 2.4 tb; ExxonMobil[9] of 3.2 tb; United States Geological Survey (USGS)[10] of 3 tb. The five-year study conducted by USGS and released in 2000 appears to be one of the most reliable assessments of the world crude oil and natural gas technically recoverable resources.

Petroleum resources are identified and assessed using geologic analysis with probabilistic methodology. The study estimates undiscovered conventional resources at 782 bb; reserve growth at 688 bb; remaining reserves at 891 bb and cumulative production at 710 bb (Ahlbrandt 2000). Given these figures about reserves, consumption data and rate of growth of both consumption and recoverability, there are roughly enough oil reserves to sustain the world needs for the next 70 years.

The beginning of the offshore oil and gas activity can be traced back to 1923, when a derrick mounted on a wooden platform started to drill for oil in the shallow waters of the Mississippi delta in the state of Louisiana. However, the first successful commercial development took place 12 nautical miles off the coast of Morgan City, Louisiana, in 1947, when Kerr-Mcgee, Stanolind and Phillips Petroleum Company drilled the Kermac 16 in 6 m of water in the Ship Shoal Block 32 field (EIA 1999). Since that historic event,[11] offshore petroleum production has constantly increased. During the last decades, production offshore has grown faster than onshore. In the 1990s about 80 percent of the petroleum production growth in non-OPEC countries came from offshore fields.

Offshore development regions are extremely widely distributed—from Alaska to Tierra del Fuego; from the South China Sea to the Gulf of Mexico and the North Sea. More than twenty major offshore development regions are operational today, with 126 ongoing offshore projects all around the world.[12] It is evident that the oceans are playing an increasing role in the world's supply of conventional energy, with more than 8,300 offshore oil and gas installations operational worldwide and a supply of 645 mobile offshore drilling rigs and 288 platforms, offshore oil and gas productions contribute respectively around 35 percent and 25 percent of the world's supply.

A study by the energy sector analyst Douglas-Westwood (2004b) predicts that offshore oil and gas production will grow from 39 mb per day recorded in 2004 to 55 mb by 2015. Production of offshore oil and gas is expected to provide by 2005 respectively around 40 percent and 35 percent of the world's supply (see Figure 3.1). Major increments to oil supply are expected to come from offshore resources located in the Caspian Basin, Latin America and deepwater West Africa. The total costs to explore for, develop and operate offshore oil and gas fields—presently some USD 111

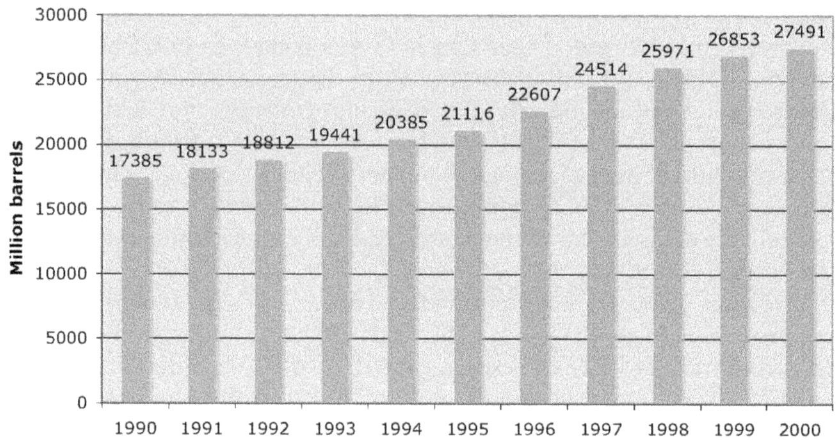

Figure 3.1 Total offshore oil production, based on Offshore Oil Prospect to 2000 © OECD/IEA, 1996, Table 2, p. 31.

billion per year—are expected to reach USD 1,440 billion over the next decade.

The success of offshore oil production has occurred despite the extremely large investment required for its development. Due to logistic problems related to the marine environment, drilling a well offshore is much more expensive than onshore. The cost of an offshore well depends mainly on the distance of the location and on the water depth, with cost ranging between USD 35–100 million. Deepwater projects have associated investment costs that may exceed USD 1 billion (EIA 1999). The offshore oil and gas industry has also a considerable impact on other sectors of the economy, including boatbuilding and metal construction, marine technologies not oil-specific, communication equipment, transportation and other water and underwater vehicles and professional services. Within the offshore drilling sector, 42 percent of the expenditure is related to engineering services, 25 percent to support and 7 percent to geo-science (Douglas-Westwood 2006).

A study surveying the worldwide service vessel fleet counted that 1,028 offshore vessels (see Table 3.1) were operational worldwide in 2003, including anchor handling tugs, anchor handling tug supplies, multi-functional support vessels, platform supply vessels and general supply vessels (Kammerzell 2003). According to a research conducted by OECD and IEA there are three main factors in the strong growth of offshore petroleum production (OECD-IEA 1996).

1. The application of new technology for gathering and processing geological information. Advanced techniques in both acoustic sensing and computer data processing have lowered the costs of finding hydrocarbons, of developing fields and improved well performance.

Table 3.1 International Offshore Rigs

February 2006	*US Gulf of Mexico*		*Europe/Mediterranean*		*Worldwide*	
	Mobile	*Platform*	*Mobile*	*Platform*	*Mobile*	*Platform*
Total rigs in fleet	156	55	101	109	645	288
Contracted rigs	129	28	98	107	580	232
Rigs utilization (%)	82.7	50.9	97.0	98.2	89.9	80.6

Well performance is of utmost importance to ensure an economically viable offshore production because it affects both recovery volumes and the speed at which volumes are produced. An accelerated production in turn determines the revenue of the project, since an increased flow rate per well reduces the number of production wells necessary to have the same level of production. The associated drilling and completion costs decrease accordingly.

2. Reductions in cost management. The new organizational approach moves projects more efficiently from the exploration phase to the final decommission. The resulting reduction in project development time contributes to a more cost-effective management of all oil and gas activities. The oil industry is frequently co-operative and multinational, in the sense that exploration and exploitation are often carried out by joint venture corporations made up of companies from different countries. Exploration efforts are cyclical; they intensify during periods of relatively high oil prices and contract during drops in the price of oil.
3. Changes in fiscal regimes.

OFFSHORE OIL TECHNOLOGY

The technology associated with offshore oil extraction has been developed quite recently, but at a very quick rate. The oil and gas industry is presently one of the most technologically advanced industries in the world. Many high-tech developments have occurred in state oil companies, especially in Europe. The general trend of exploration and production activities has been to gradually move offshore exploration further out into deeper waters and a harsher operating environment. In the late 1970s the deepest production was about 300 m. The Roncador field in the Campos Basin, Brazil, established in 1999 an offshore production record depth at 1,800 m. After four years, a well was drilled in the Gulf of Mexico by Chevron-Texaco-Toledo in 3,051 m of water. It is worth reminding that after having penetrated the sea floor, these well bores extend hundreds of meters below the ocean floor. As a result of equipment innovations made from titanium alloys[13]—that can withstand the high temperatures and pressure found at very deep wells—the oil and gas industry can now develop fields where temperature is around 200° C and pressures about 1,100 bar. According to Douglas-Westwood (2004b) in 2015 nearly 25 percent of offshore oil will be produced from deepwater,[14] compared to just the 10 percent share of the production in 2004.[15] The technological progress has also benefited the working conditions of the rig crews, which is safer especially for personnel operating at deepwater wells.

A number of different platforms are employed in offshore oil and gas operations (see Figure 3.2), each of which has distinctive characteristics and operating depths. The major distinction is whether a platform is free-floating or bottom-supported. The main drilling unit types are described below (EIA 1998).

Fixed platforms

They consist of a jacket and a deck. The jacket is a steel structure similar to a crane that stands on pillars imbedded in the ocean floor. The deck is placed on top of it to host the crew and the production facilities. Fixed platforms represent the first generation of drilling units and are suited to drill only in shallow waters of a maximum depth of about 500 m. They need to be decommissioned at the end of the oil field production (usually destroyed with explosives) and therefore also cause some environmental problems.

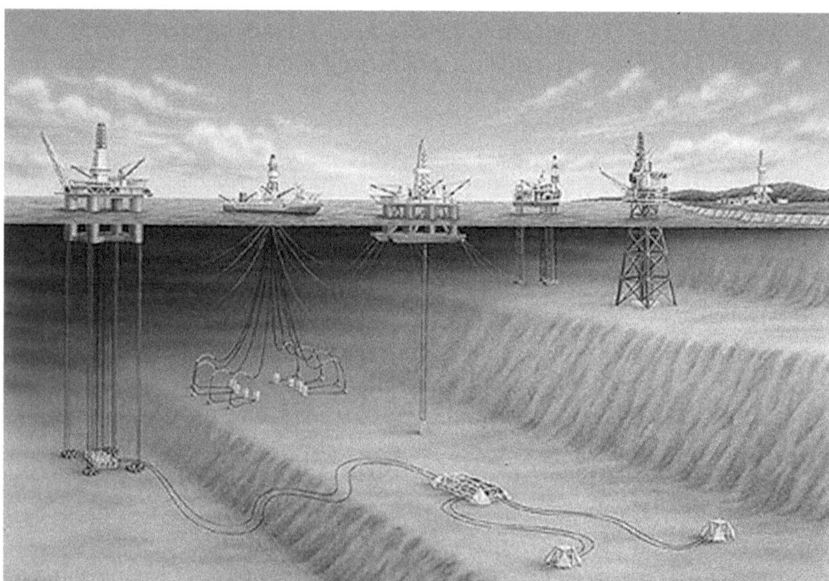

Figure 3.2 Offshore Drilling Units. From right: Jacket barge, Jack-up rig, Semi-submersible unit, Drill ship and Tension leg platform. Source: BC Offshore Oil & Gas Technology Update (2001).

Semi-fixed platforms

Instead of being embedded in the seafloor or floating over the drilling location, semi-fixed platforms are supported by long retractable steel legs lowered on to the seabed to raise the rig out of the water. The most used semi-fixed platforms are Jacket barges and Jack-up rigs. Jacket barges are usually employed in shallow waters of maximum 50 m. Jack-up rigs, although generally used in shallow waters up to 150 m, have been utilized in water depths of up to 1,000 m, however, their cost rises rapidly at increasing depths. Depth permitting, these are the most frequently used drilling units because of their excellent stability under various weather conditions.

Self-propelled drilling units

There are two main types of self-propelled units: semi-submersible rigs; and barges. A semi-submersible unit is suspended above the surface and supported by buoyant pontoons and columns that, when flooded with seawater, cause the pontoons to submerge to the predetermined depth. Although smaller than other platforms, these units still have a considerable dimension and can host over one hundred people onboard. For instance, the Troll A platform gas rig, owned by Statoil, is a structure 472 m high with 369 m laying below the surface. Being self-propelled, once they have completed their task in an oil field these units can move to other locations. State-of-the-art semi-submersible rigs can drill in water depths of over 1,500 m.

Floating platforms and drill ships

Floating platform units include a number of different floaters such as Submersibles, Spars, Truss-spars and Single column floaters. However, the most important floating platforms are Tension leg platform and Floating Production Storage and Offloading (FPSO) units. Tension leg platforms can have different size and are deployed in water depths approaching 1,300 m. They are composed by a floating structure held in place by vertical tendons connected to the sea floor by pile-secured templates. FPSO is a production system that employs a large tanker-type moored vessel. This system has been devised to process and store the oil, which is periodically downloaded into a smaller shuttle tanker. Drill ships are just vessels that have been fitted with a drilling apparatus and donned of a DGPS to maintain a fixed position over the well. Although they are not as stable as semi-submersibles drilling units, a series of small thrusters combined with DGPS technology allow drill ships to remain essentially stable despite winds and currents, shifting less than 15 m in any direction. This considerable stability allows the drill ship to operate in water depths of over 2000 m, in areas without pipeline infrastructure and in most weather conditions. Unlike semi-submersibles and Jack-ups, drill ships do not require to be towed to the drilling location and do not have any associated decommissioning cost. According to a 2004 forecast, 110 floating production systems are to be installed over the next five years and the capital expenditure required to complete these orders is expected to be USD 33 billion (Douglas-Westwood 2004b).

Offshore oil and gas activity can be divided into four main phases:

1. Exploration
2. Development
3. Production
4. Decommissioning

Exploration

The exploration phase can be further divided into two stages: pre-exploration; and exploration. Pre-exploration operations mainly consist of: (A) collecting basic data on ecological, scientific, social and economic issues; and (b) assessing both the jurisdiction and the environmental aspects and requirements.

(A) Regarding social-economic issues, oil companies often try to maximize the benefits for the local coastal area and the resident communities. Benefits that usually spin-off from oil and gas activities are opportunities such as local access to natural gas supply, the development of processing facilities and businesses that service the industry. Even if in a short-term perspective, usually it is the development activity that supports the highest number of local jobs. On the other hand, production activity offers fewer opportunities of local employment but does so on a long-time basis.

(B) Offshore oil production is usually carried out under a system of leases and licensing of the concerned area for a specified amount of time. Generally speaking the life of hydrocarbon reservoirs is uncertain, usually ranging from 10 to 40 years. The specific characteristics of the leasing agreement vary from country to country, however, it generally takes the form of licenses from the government to the commercial firm that wants to explore for oil. For the exploitation of the field another license is usually required.

Oil exploitation is a long-term activity the effects of which extends and cumulates over time. Assessing the environmental impact and the possible conflicts with other marine activities are critical for the development of the oil field and the sustainability of the project. The main environmental short-term issue related to the exploration activity concerns the impacts of seismic testing on fisheries and marine mammals, while the long-term effects concern the physical, chemical and biological characteristics of the coastal ecosystem. Regarding conflicts, although the classic offshore conflict is oil vs. fisheries, in some locations the emerging importance of tourism and its related recreational activities calls for special attention to the impact on this sector. Another possible conflict can arouse with the expanding industry of underwater cabling.

The exploration phase aims to find hydrocarbon reservoirs and determine the feasibility of the exploitation process. This task has a high degree of uncertainty, since there is no guarantee that exploration will lead to development and production. The exploration phase may continue on and off for years without developing the field. Exploration is typically a short-term work conducted by specific professionals. It requires the employment of expensive and mobile technical equipment, including seismic vessels, drilling rigs, support vessels and even helicopters. Usually, this equipment is owned by multinational enterprises and operated by specialized freelancers that undertake the exploration activity for oil companies on a contractual basis (BC Offshore Oil 2001). Exploration consists of two primary activities: geological surveying and data interpretation; and exploratory drilling. Environmental assessment for seismic and exploration drilling programs are always required.

Geological surveying

To find oil or gas, geologists first map the seafloor characteristics looking for areas where it is geologically likely that a hydrocarbon reservoir can be found—i.e. rock formations that are commonly associated with natural gas and petroleum. A marine geological survey includes the use of satellite imaging, sensitive gravity meters, sensitive magnetometers, 'sniffers' and marine seismic surveys. Gravity meters measure changes in the Earth's gravitational field that could indicate flowing oil. Magnetometers measure changes in the Earth's magnetic field caused by flowing oil. Sniffers detect the smell of hydrocarbons, while marine seismic surveys employ seismic energy to probe beneath the seafloor.

Present offshore oil exploration techniques have in average a success rate of about 10 percent. The most important element to detect an underwater hydrocarbon reservoir is by far the marine seismic survey.

In marine seismic surveying, geophysicists artificially generate shock waves that are reflected back to the surface from the rock formations located under the seabed. The result is an image of the subsoil created by a computer that must be analyzed and interpreted by specialists. The main aim of this analysis is to detect 'areas of closure', which are areas that often act as hydrocarbon traps.

Underwater shock waves can be created in two ways: (1) by a compressed-air gun—which shoots pulses of air into the water; or (2) by explosives—which are thrown overboard and detonated.

A reflection seismic survey typically involves generating hundreds to tens of thousands shots at different locations in the survey area. Shock waves travel in the water and beneath the seabed and are then reflected back by the various rock layers. The reflection waves travel back at different speeds depending upon the density of the rock layers through which they pass. The reflection waves are detected over water by a vibration detector called hydrophone. As already mentioned in the previous chapter, the hydrophone is a receiver that transforms sound movements in the water into electrical voltage. Seismologists finally interpret the hydrophone's reading—i.e. seismograms or traces, which are graphical time series of the Earth movements—looking for signs of oil and gas areas of closure.

Marine seismic surveys are generally conducted with large ships and using multiple air gun arrays. Air guns are deployed behind the vessel to generate seismic signals by forcing highly pressurized air into the water. Hydrophone receivers are towed behind the ship in long streamers several kilometres long. Up to 20 streamers can be deployed from a vessel, allowing the acquisition of large quantity of data at a time. A recent advance in marine seismic surveying technology is the use of multiple vertical cables. Each cable is anchored at the seabed and contains multiple receivers, thus acquiring data at multiple depths and facilitating reading and interpretation. Marine receivers are composed of a group of 5 to 50 piezoelectric hydrophones,[16] which respond to changes in water pressure producing a single seismogram. This equipment can probe subsurface depths of up to 10 km. Computer assisted marine seismic surveying can acquire and image data in 2D, 3D and 4D. In a conventional 2D marine seismic survey a vessel usually tows an array of 12–70 air guns at depths of 4–8 m, with air guns typically firing once every 10–15 seconds or every 25 m. (Offshore Oil and Gas Research Group 2004). Several recordings are registered along parallel lines spaced 0.5 up to 1 km in two orthogonal directions. 2D imaging requires only several hundred data points. Although 2D seismic imaging is less complicated and less detailed than 3D imaging, 2D imaging technique is an extension of the 3D technique. Since it is simpler than 3D, 2D imaging is much cheaper, easier and quick to perform. Usually several surveys are needed to accurately define geologically an area, each of which typically lasts a few weeks.

In the last decades marine seismic survey technology has developed less invasive techniques resulting in computer enhanced 3D and 4D data collection and seismic imaging. Imaging in 3D requires data to be collected from several thousand locations and the survey is therefore developed towing multiple rows of source a cost up to about USD 1 million per 40 square nautical miles (Offshore Oil and Gas Research Group 2004). Stationary sensors installed on the seabed are used to generate 4D seismic images that show changes over time in the structure and properties of the underground rock formations. Typically, marine seismic data are collected 3D by large specialized contracting companies. Once the data are acquired, other specialized contractors electronically process and interpret this impressive amount of data.

Advances in data modelling and processors power, enable today the creation of more accurate and detailed geological profiles. This progress has improved the success rate for exploratory wells and has therefore lowered both cost and environmental impact. Satellite RADAR imaging, aeromagnetic surveys and gravity surveys are also used in helping identify characteristics and location of hydrocarbon reservoirs. Gravity and magnetic surveys detect anomalies from the area average value, improving the overall picture of the sedimentary basin. High-resolution aeromagnetic surveys, taken at 60–90 m altitude, provide additional details, such as fault trace and near volcanic rock (BC Offshore Oil & Gas Technology Update 2001). After the geological survey of the underwater area has taken place, exploratory drilling occurs only if the evaluation of the economic feasibility has produced good results.

Exploratory drilling

Once the underlying geologic strata have appeared promising and the potential for hydrocarbon deposits in the surveying area have been assessed with a certain degree of confidence, drilling an exploratory well into the geological structure to discover what it contains is the next step. This process can last up to 3–4 weeks. Wells are drilled during the exploration phase both to confirm geological data and to delineate an estimate of the size and quality of the reserves. The exploratory well allows geologists to study in detail the composition of the underground rock layers by examining the drill cuttings and fluids.

Furthermore, a series of tests called 'logging' is performed during or after the drilling process in order to monitor the rock characteristics and response. By integrating data obtained from the different types of tests, it is possible to create a detailed visualization of the underground formation. The exact placement of the spot to drill depends on several factors, such as the nature of the rock formation, the characteristics of the underwater subsoil geology and the depth and size of the target reservoir. If the drilled well makes contact with the hydrocarbon and the estimation of the marketable quantity of oil and/or gas was correct, the well is developed to allow for the extraction (productive well), otherwise the development activity does not proceed (dry well).

As previously mentioned, drilling is usually conducted from a Jack-up rig if the reservoir is located in shallow waters (less than 100 m) or from a semi-submersible rig or a drill ship if located in deeper waters (see Figure 3.3). The exploratory well is drilled in a series of steps of decreasing size within the increasing depth. During the exploratory drilling, additional data are collected in order to optimize the selection of the drilling rig and the placement of the drill holes and of the related seabed installations. This technique, called Measurement While Drilling (MWD) includes Vertical Seismic Profiling (VSP), Magnetic Resonant Imaging (MRI), gravity and

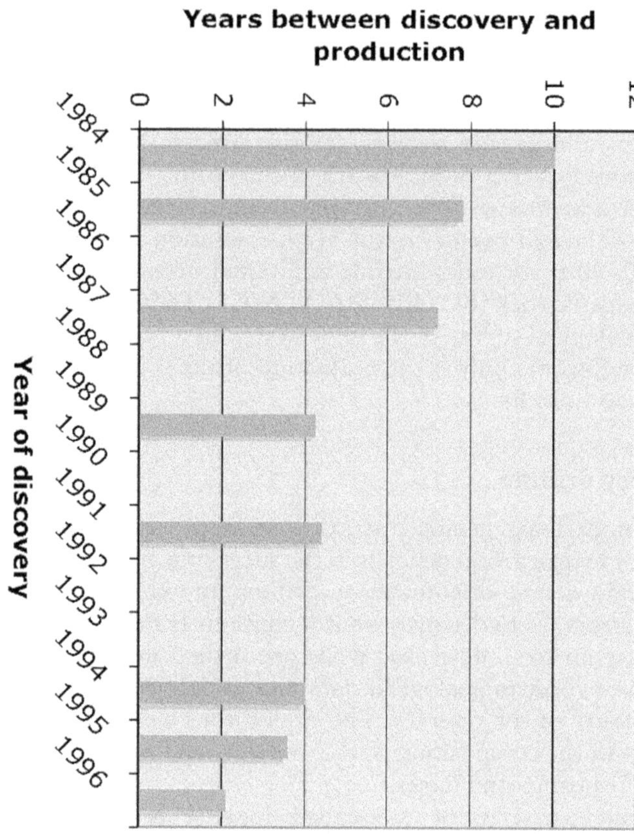

Figure 3.3 Cycle Time for Deep-Water Projects. Source: EIA (1999).

resilience measurements, Neutron Porosity Measurement, and geo-steering and pressure detection capabilities (BC Offshore Oil & Gas Technology Update 2001). The technique used to drill an exploratory well and a production well is almost similar and is described in the next section.

Development

Once a commercial field has been identified and surveyed, the exploratory well is drilled and analyzed and the reserves assessed, the oil company may proceed to develop the field. Development activity includes getting regulatory approvals for the production, planning the stages of development of the field, analyzing and selecting drilling prospects, constructing and installing production and transportation systems and, of course, drilling production wells. Usually interpreting exploration results and getting the regulatory approvals take more than the time to implement the seismic survey and drill the wells. Technological advances in drilling rigs are progressively limiting the use of fixed platforms, which involve numerous legal and environmental requirements. Consequently, the use of floating production systems such as FPSO is an increasing trend in offshore oil production, even for large-scale construction projects. The evolution from the exploration to the production phase normally occurs over a few years.

Production

The production phase involves extracting, developing and monitoring production and operating facilities, evaluating reservoir's performance and transporting hydrocarbons. As already mentioned, production from each single field can last from 10 to 40 years, according to the reservoir's consistency, the extraction rates and to petroleum prices. The process of extracting oil offshore starts with the drilling of a well to access the reservoir. All types of offshore rigs house the necessary equipment to drill and extract oil and/or natural gas from underwater reservoirs. The drilling rig is the basic gear used to dig a well into the seabed.

A drilling rig is a complex apparatus composed by many parts. Basically, the drilling assembly consists of a series of connected hollow cylindrical steel pipes that form a long flexible drill-string designed to make a borehole linking the reservoir to the surface. The drill-string is driven in a rotational motion from the top end through a gearbox—called 'top-drive' or 'Kelly drive'—connected with an electric motor. The drill-string is typically made up of three components:

1. the drill pipe, made up of connected sections of a length of about 10 m each;
2. the transition pipe, needed to connect the drill pipe to the surface motor;
3. the Bottom Hole Assembly (BHA), which is the short and heavy terminal part made up of a drill bit, the drill collars and the stabilizers.

The drill bit is the cutting device that penetrates the rock. Depending on the drilling task and the characteristics of the rock formation, the drill bit can have different shapes, such as roller or fixed cutter blades. It can be made of different materials, such as tungsten carbide steel, diamond, or diamond-impregnated bits. The drill collars are large and heavy pipes employed to place weight on the drill bit. The stabilizers keep the drilling assembly centred in the hole. Other components, such as the downhole motor, the stabilizing fins (designed to minimize lateral motion during drilling) and MWD and LWD (Logging While Drilling) tools are usually part of the BHA (Wu and Chen 2006).

Oil and gas wells are constructed in stages, with a system (called 'casing') of multiple layers of metal pipes slightly smaller than the size of the hole.

Wells are drilled from top to bottom in a series of progressively smaller-diameter intervals of casing. The diameter of the hole varies according to the depth of the well, being larger (about 1 m) for shallow waters and smaller (about 60 cm) for deeper water. After the first hole is drilled at a certain depth, the casing is fixed into the hole. The casing is then bonded and secured to the hole by pumping cement through the space between the outside of the casing and the hole. The casing provides structural integrity to the newly drilled well bore, preventing the hole from collapsing. It also isolates potentially dangerous high-pressure zones from each other and from the surface. Once the dangerous zones are safely isolated and the formation protected by the casing, the well can be drilled deeper with a smaller diameter bit, which is also cased with a smaller size casing and then cemented. This process is repeated several times, wells often having two to five sets of subsequently smaller holes drilled inside one another. The number of the nested holes depends on the characteristics of the well, such as the total depth and the strength and sensitivity of the formation through which the well passes, and on regulatory requirements. The excess in the diameter of the casing over the diameter of the drill string is called annulus. This ring-shaped void is necessary for the conduction of the drilling fluids or muds.

During the process of drilling, a pressurized fluid called drilling fluid or drilling mud is continuously pumped through a circulation system down the hollow centre of the drill string, exits out of holes located in the drill-bit and then returns back to the surface via the annulus. The drilling fluid or mud is a lubricant used for three main purposes:

1. To cool and lubricate the drill-bit;
2. To remove the cuttings produced by the bit at the bottom of the hole and carrying them to the surface;
3. To maintain hydrostatic equilibrium so that fluids and gas present in the reservoir do not enter the well bore causing the well to flow, kick or even blow out.

There are different kinds of drilling fluids. Usually the classification is based on the composition of the base fluid:

1. Water-based Drilling Muds (WBMs);
2. Oil-based Drilling Muds (OBMs), composed by diesel or mineral oil;
3. Synthetic-based Drilling Muds (SBMs).

Drilling fluids contain a variety of mineral and chemical additives intended to give them the required properties to serve specific functions. Additives to drilling muds are numerous. The most important ones are weighting agents such as barite, hematite and calcium carbonate to balance the pressure in the well, thickener agents to enhance the viscosity of the fluid in order to remove rock chippings, and deflocculant agents to reduce viscosity. Traditionally, WBMs have been used extensively for two reasons. First, because they are cheaper and second because, as long as they meet some environmental standards, can be disposed of directly at sea in most offshore locations of the world. However, for tricky drilling situations, such as deep wells and horizontal and extended-reach drilling, WBMs do not offer good drilling performance and therefore OBMs must be employed.

Since the 1990s, there has been an extensive use of SBMs because they match the good drilling properties of the OBMs but do not contain polluting agents such as polynuclear aromatic hydrocarbons. SBMs have a lower degree of toxicity, faster biodegradability and lower bioaccumulation potential than OBMs. SBMs generate a lower volume of drill cuttings but they are more expansive than WBMs and therefore they are recycled to the possible extent (Veil and Dusseault 2003).

The cuttings produced by the drill-bit are taken to the surface by the drilling mud through the annulus. Once onboard the drilling unit, cuttings must be removed from the drilling mud in order to recycle the fluid. The shale shaker is the device that separates and removes the rock cuttings from the drilling fluid. Its main component is a mesh screen that vibrates while the drilling fluid flows on top of it. The liquid of the mud and solids smaller than the wire mesh pass through the screen and go back to the mud pit—where the drilling mud is mixed and recycled in order to be re-circulated. The larger solids are retained on the screen and a shale slide conveys them to the cuttings pit. Often two series of shale shakers are used: the primary shakers—which have a coarse mesh screen in order to remove only the larger cuttings—and the secondary shakers, which have a finer mesh screen to remove smaller solid particles. When the cuttings coming from the well begin to contain oil sand from the reservoir, the planned final depth may have been reached. At this stage, the drilling apparatus is removed from the hole and several tests are performed to confirm the finding. The main tests used today are:

1. Well logging, through which measurements of the rock formations are taken by lowering electrical and gas sensors into the well hole;
2. Drill-stem testing, which reveals—by measuring pressure—if the reservoir has been reached;
3. Core samples, to examine the characteristics of the reservoir.

Once the planned depth has been reached and the performed tests are positive, the production equipment is set up to extract the oil from the well. However, the well must first be completed in order to allow the liquid to flow into the casing in a controlled manner. For this purpose a path must be made for the oil to flow from the reservoir into the well bore. Therefore the casing that covers the reservoir section must be perforated. Lowering in the well a perforating gun with explosive charges is the means to make holes in the casing. When the reservoir is made by limestone rock, acid is pumped down the well and out the perforations. The acid creates small channels in the limestone that lead oil into the well. When the reservoir is made by sandstone rock, a special fluid containing proppants (such as sand, resin-coated sand, high-strength ceramic materials like sintered bauxite, etc) is pumped down the well and out the perforations. Pressure coming from this fluid provokes small fractures in the sandstone allowing the oil to flow into the well. The proppants are needed to hold the fractures open. Once the casing has been perforated and the acid pumped down, a small-diameter pipe called "tubing" is run into the hole as a conduit for oil and gas to flow up the well. Then a device called "packer" is run down the outside of the tubing and expanded to form a seal around the outside of the tubing. Finally, two devices connected to the wellhead complete the extraction system: a "Christmas tree"; and a blowout preventer (BOP).

The Christmas tree is a diverter system—i.e. a structure composed by many valves—connected to the top of the tubing and cemented to the top of the casing. It serves to regulate the pressure, to control the flow and to allow access to the well bore in case further completion works need to be done. From the outlet valve of the Christmas tree, the oil flow can be connected to a distribution network of pipelines and tanks. The BOP is a large high-pressure valve that seals the high-pressure drill lines and helps—when necessary—to relieve pressure in order to prevent a blowout. BOPs come in a variety of styles, sizes and pressure ratings. Since BOPs are critically important to the safety of both the crew and the drilling rig and the well bore, they are inspected, tested and refurbished at regular intervals. BOP testing varies from daily function testing on critical wells to monthly or less frequent testing for more stable wells. Once a well is finally equipped with the mentioned devices, compressors are used to pressurize the well, thus allowing the crude oil to exit the well through a vertical shaft. The compressed oil is then transported to processing sites and finally to a storage tank (Wu and Chen 2006).

When free-floating platforms or drill ships are employed to extract oil, the critical component is a flexible hollow pipe called marine riser. The latter provides the transferring of fluids and power between sub-sea units and the topside floating platform or drill ship. The drill-string is connected to the floating rig passing down the marine riser, which is fixed both to the floating rig and the seabed wellhead. It protects the rotating drill-string from the sea, keeps the drill cutting debris from escaping into the sea and offers a conduit for the transit of oil and gas. The integrity of the marine riser under a variety of harsh marine conditions is crucial. These pipes are excited by both strong external forces due to currents, waves, vessel motion and seabed contact, and by internal forces due to internal flows. Therefore, they must have high structural axial stiffness and relatively low structural bending stiffness to allow large deflections to accommodate high internal and external pressures. The marine riser is kept in a state of mechanical tension to minimize its lateral motion. Due to the great length involved in deep-sea operations and the strain to which the marine riser undergoes, a flexible pipe is essential. When operating in deep waters, wellheads and flow-lines are installed and maintained by ROVs. ROVs are widely used in the offshore oil and gas industry for the installation and maintenance of a number of sub-sea structures—e.g. operations such as deployment of marine risers and control umbilical, and pipeline inspections.

An advanced technique presently in use in the offshore oil and gas industry is directional drilling. Traditionally, oil wells were drilled vertically and completed at the required depth to intersect a single reservoir. Consequently, a well was necessary for each reservoir. Since the late 1970s advances in drilling technology allow much more flexibility, permitting a greater production with less drilling and therefore a less amount of cuttings to dispose of. Technologies such as steerable downhole drilling motors and MWD tools have now become customary for the offshore oil and gas industry. Downhole drilling motors permit the bit to be rotated on the bottom of the hole, thus allowing drilling to proceed at angles off of vertical. A recent advance of this technique has been the development of a range of rotary steerable tools allowing 3D control of the bit without shutting down the drill-string rotation. In addition, the development of mud pulse telemetry permits to gather down-hole data directly at the surface while drilling. These new advances unfolded new possibilities for improving production, paving the way to additional technological developments in drilling, such as:

1. Horizontal drilling, which is primarily used when a reservoir is not thick enough to be drilled vertically but extends over a large lateral area. Horizontal drilling allows the well to be drilled and completed in a thin horizontal layer. Before the introduction of this drilling technique the exploitation of such reservoirs was either not economically viable or required multiple wells to extract the hydrocarbons;
2. Extended-reach drilling, which is a technique used in situations where it is not feasible or too expensive to drill the well directly from above the target reservoir. With extended-reach drilling many wells in the same oil field can be completed from just a single location, reaching reservoirs even several kilometres away from the drilling place. This technology also avoids the environmental impact caused by several surface support structures;
3. Multiple drilling, which is a technique used when the reservoir is made-up of small multiple zones, or zones at several different depths. Using directional drilling technology, lateral well bores can be drilled from the main vertical well bore, which extends to reach each target zone. Once the main vertical well bore is drilled, several smaller-diameter lateral perforations follow;

4. Slim-hole drilling, which is a technology that—even if available since the 1950s—has only recently been refined and extensively used. According to the US Department of Environment, slim-hole wells are defined as wells in which at least 90 percent of the hole is drilled with a bit of six inches (15.24 cm) or less in diameter. Conventional wells typically use drill bits as large as 12.25 inches (31.11 cm) in diameter. Cutting-edge technology in this field is now micro-hole drilling, in which the shaft is less than 5 cm in diameter, resulting in a smaller environmental footprint and less surface disturbance;
5. Coiled tubing drilling, which is a drilling technique that does not connect individual sections of the drill pipe as in the drilling processes described so far. In its place a continuous length of tubing is fed down the borehole. The coiled tubing has a smaller diameter than traditional drill pipes, thus generating a smaller volume of cuttings and fewer disturbances to the surrounding environment.

As a result of these technological advances, drilling rigs can more easily turn the well bore to reach targets horizontally from the position of the wellhead. The dimension of the covered area can reach several hundred square kilometres. Grouping together more wellheads (up to about 60 satellite wells) on one surface location gives several advantages, such as fewer rig moves, less surface disturbance and an easier and cheaper way to complete and produce the wells. This technique is accomplished by using a manifold, which is a device that ties together produced fluid flowing from satellite wells and pumps it up to the storage vessel. The choice of the production system varies accordingly to the specific characteristics of the oil field and its reservoirs, such as water depth, environmental conditions, the physical and chemical properties of the oil and/or gas, and the size of the field. In the past, fixed platforms made up of steel or concrete and anchored to the sea floor were mainly employed. However, the past decade has seen an increasing use of floating and sub-sea systems, including FPSO vessels, tension leg and spar platforms. The main issue for the offshore oil technology is that today it is no longer necessary to build permanent structures over each and every hydrocarbon reservoir. Therefore, instead of fixed-leg platforms—and especially in the case of deepwater projects—offshore facilities float on tethered platforms or are directly located onboard drill ships.

The transportation of hydrocarbons onshore is accomplished using tankers and/or pipelines. The choice depends on a number of factors, including the location, the physical and chemical characteristics of the oil and/or gas, the seabed topography and geology, and the ease of access to the pipeline infrastructure. Gas is typically transported from offshore sites via subsea pipelines, while oil and gas condensates are typically transported via tanker.

Decommissioning

Decommissioning implies two main operations: well capping, and removing some or all the equipment used for production. Decommissioning offshore oil and gas facilities requires a few months. Explosives are typically used to remove structures, however, in some cases submarine structures are left in place. Structures weighing up to 4,000 tons in areas with depths less than 100 m must be removed completely, while in deeper waters only the upper parts from the surface down to 55 m depth are usually removed. The removed parts can be either transported onshore for disposal or buried at sea. Explosive charges are used in 90 percent of the cases, since removal of fixed platforms is almost impossible without using explosives (Patin 1999). Today the possibility of alternative use of abandoned structures is almost always considered. From the conservation fisheries perspective, converting the fixed marine structures into artificial reefs—thus increasing the bioproductivity of the area—is usually welcome. On the other hand, structures or parts left on the sea bottom may cause vessel and gear damages when fishing and therefore are opposed by fishers. An efficient way of decommissioning offshore platforms would be to convert them into scientific laboratories for *in situ* studies.

ENVIRONMENTAL IMPACT

The main anthropogenic flows of oil pollution into the marine environment come from (Patin 1999):

1. Land-based sources such as refineries, municipal wastes and river runoff (about 50 percent);
2. Transportation activity (about 25 percent);
3. Atmospheric fallout (13 percent);
4. Natural seeps (11 percent); and
5. Offshore production discharge (2 percent).

Even though improvements in technologies and management practices could help in reducing the magnitude of the environmental impact of the offshore oil and gas activity on the marine ecosystem, there is still the possibility of negative effects at all phases of the industrial activity. However, given its utmost importance for the world economy, hydrocarbon exploitation cannot stop for the sake of its negative environmental effects or the risk of possible accidents. The benefits coming from oil and gas are much bigger than the environmental damage caused by the operations necessary to produce them. Until new substitutes are found, hydrocarbon production must increase to sustain and fuel the growing world economy.

Due to the fact that detecting changes and understanding cause-effect relationships in the marine ecosystem is complicated, much of the research on the environmental impact of offshore oil and gas focuses on specific discharges, individual fish species or short-run effect. There is, therefore, little information regarding the overall impact and cumulative effect on the marine environment. In particular, there is a knowledge gap regarding the impact over valuable fish species. This prevents scientists to quantify the tradeoff between fisheries and oil and gas. However, even if there is no data to quantify it, it is evident that the economic tradeoff between hydrocarbons and fisheries can be easily solved if the richer hydrocarbon industry would repay the fishing industry for the losses caused by oil and gas activities.

Although there is uncertainty about the nature and magnitude of environmental impact provoked by the offshore oil and gas industry, it is known that it depends on different factors such as: the national regulatory framework governing the industrial practices; the technology used; the scale of the project; and the characteristics of the location. In order to have a clear picture, it is useful to divide polluting phenomena according to their (A) regular or (B) accidental occurrence.

In addition, the occurrence of the pollution can be linked to the phase of the industrial activity during which it happens and to the spatial (local or widespread) and temporal (short or persistent) dimension of its effects. However, predicting the severity of an environmental impact is generally a tricky task because each impact is unique to each ecosystem. Moreover, the probability that can be assigned to accidental events such as a major oil spill caused by a tanker accident is just guesswork.

POLLUTION OF REGULAR OCCURRENCE

During exploration activity

The environmental impact on the marine ecosystem starts with the exploration activity. Sonar mapping itself produces sound waves that have unknown potential affects on marine biota. Even if the exact impact of man's generated sound on marine life is poorly understood, according to research seismic surveys generate loud underwater sounds (up to 255 dB) that can be detected in water hundreds of kilometres away from the source (Offshore Oil and Gas Research Group 2004). These sounds can cause lethal impact to nearby marine biota (within about 6 m of air-gun shot) and sub-lethal impacts to distant ones, including physiological (mainly hearing) damage and changes in feeding, reproductive and other fish behaviours. Research found huge falls in catches after seismic surveys. This environmental impact has usually a local extent and mainly provokes short temporal effects.

During development activity

Once the development activity has started, physical changes occur in the drilling area. The seabed habitat is modified by the installation of offshore structures and marine populations are consequently affected by these changes. These physical changes may have both negative and positive consequences on marine life. Some local species may be negatively affected, while other species may colonize the area. The area where offshore oil and gas activities take place is commonly interdicted from fishing and since fish usually become used to the surrounding noise generated by the drilling equipment, sometimes the drilling area becomes a refuge housing a large number of fish. This kind of environmental impact usually has a local extent but provokes persistent effects.

During production activity

The main polluting activities of regular occurrence connected with the production phase are drilling wastes discharge and small spills. These effluences adversely impact marine life and habitat provoking chronic local pollution on the surface, the water column and the seabed. In the process of well drilling, gaseous, fluid and solid pollutants are discharged in the sea. Drilling wastes are primarily composed of cuttings and muds, which are in varying degrees toxic to marine biota. Drilling wastes alter the local habitat by burying the original sea floor, thus affecting in particular local benthic communities. In addition, drilling discharge increases the water column turbidity, which in turn reduces light penetration affecting various marine organisms.

One of the main problems in studying the effects of drilling waste discharge is the identification of the additives used in offshore chemicals. Drilling fluids are made from a range of over 1,000 products and producers are reluctant to communicate their formula. As a result, a reliable assessment of the entire extent of the environmental damage provoked by drilling wastes has not been quantified yet. However, it is undoubtedly known that discharge of pollutants at sea threaten marine life, especially sessile and benthic organisms. For example the so-called "produced water" or "production water", which naturally occurs within oil and gas reservoirs,[17] is extracted together with hydrocarbons, treated in order to remove as much oil as possible and then discharged into the sea or re-injected into the wells. Typical constituents of produced water include hydrocarbons, greases, barium, heavy metals and various chemical additives. Produced water represents the greatest amount of discharge from offshore platforms: up to 3,000 tons per platform per day, the oil content varying within the range of 23–37 mg/l (Offshore Oil and Gas Research Group 2004).

In the early stage of offshore oil and gas, drilling wastes were usually discharged from the platforms directly into the ocean. During the 1970s and 1980s, with the introduction of OBMs, it became evident that discharges could have adverse effects on local marine ecosystems. Today a number of jurisdictions require its disposal onshore or its reinsertion into the reservoir. There are, therefore, three main options to dispose cutting wastes: discharging; cuttings re-injection (CRI); and hauling drilling wastes onshore for disposal (recycling, landfill and/or incineration).

The solid cuttings collected in the cutting pit through the shale shakers are first analyzed, if they meet the discharge standards cuttings are usually discharged, thus avoiding the treating cost. Cuttings containing WBMs can be discharged. However, if they do not meet environmental standards for discharge they must be cleaned onboard prior to discharge. Cuttings can be processed through size dryers, squeeze presses or centrifuges. The aim is to reduce particles until dry, powdery cuttings are produced.

Cuttings containing SBMs may not be discharged untreated. CRI is a waste disposal technique that consists of injecting drilling wastes into underground formation for permanent disposal. Re-injection also helps in maintaining reservoir pressure in order to avoid crackdowns. Although there are several ways of injection, the basic one is generally called slurry injection.[18] This involves reducing drill cuttings and other oilfield wastes into small particles, mixing them with water or other liquids to slurry and finally injecting the slurry into an underground formation at pressures high enough to break the rock.

There are two main ways of injecting cuttings: annular injection; and injection into a disposal well. Annular injection brings the waste slurry into the reservoir through the annulus. Injection into the disposal well involves injecting to sections of the hole that can be either below the casing or to a perforated section of the casing. In multi-well drilling for each successive drilled well the wastes are injected into the previously drilled well (sequential annulus injection). This is usually the preferred way to dispose of cuttings because it avoids the cost of drilling a dedicated disposal well. Injection cost is about USD 500,000 per well. Most operators choose one of these options in order to avoid transportation cost onshore.

Regarding regulation and control of dumping industrial wastes at sea, in Europe the first agreement was the Convention for the Prevention of Marine Pollution by Dumping from Ships and Aircraft (called the "Oslo Convention") signed in February 1972 and entered into force in 1974. A similar document is the Convention for the Prevention of Marine Pollution from Land-Based Sources (the so-called "Paris Convention") signed in June 1974 and entered into force in 1978. It regulates the prevention of marine pollution by discharge of dangerous substances from both land-based sources and offshore platforms. This agreement requires the application of: the precautionary principle; the polluter pays principle; the best available techniques (BAT); and the best environmental practice (BEP).

The Convention for the Protection of the Marine Environment of the North-East Atlantic (a.k.a. the "OSPAR Convention")—entered into force in 1998 and replaced both the Oslo and Paris Conventions—is presently the basis for national law governing the discharge of offshore drilling wastes in the waters of the OSPAR signatory states.[19] The OSPAR Convention states that it is legal to discharge WBM cuttings in the offshore waters of the OSPAR signatory countries provided that: (A) the oil content is less than 1 percent by weight and (B) the material has passed tests to show that it will bio-degrade over a specified time and will not bio-accumulate. However, the reporting and monitoring system for discharge of offshore wastes is based almost entirely on self-reporting, therefore, all data collected are unverifiable. Moreover, unexpected visits to platforms by government inspectors are almost impossible because helicopter access to the installations is controlled by the oil industry.

POLLUTION OF ACCIDENTAL OCCURRENCE

Pollution causes by accidental occurrence is unavoidable. As oil is extracted, refined, transported, distributed or consumed, spills and other accidental releases may occur. Depending on a combination of natural, technical and technological factors, causes, scale and severity of the accidents' consequences are thus extremely variable. Usual causes of accidents include equipment failure and personnel mistakes, however, the most common accidents occur when drilling and during transportation or storage. The two main related hazards are unexpected blowouts and hydrocarbon spills. The most severe potential impact of drilling is a possible—even if remote—blowout, which can cause a major environmental damage analogous to a large oil-spill. This can happen when high-pressure reserves of hydrocarbons are drilled. The major threat of transportation activity is the spill resulting from tanker or pipeline accidents.

Oil spills

Major oil spills occur only occasionally, however, they receive considerable public attention because of the evident environmental damage. Since they mainly occur during transportation, since 1999 about two-thirds of the tankers operating worldwide have either double-hull or segregated tank arrangements. These solutions have reduced oil pollution caused by tanker transportation in the last years. Each year about 6,500 large tankers transport more than 1.2 billion tons of oil and oil products worldwide. Oil extracted offshore accounts for about 50 percent of the annual volume of oil transported by tankers. Minor oil spills receive less public attention but usually these small amounts are released continuously over long periods, creating chronic contaminations over huge areas of the marine environment. Typical chronic sources of oil spills include point sources, such as natural seeps, leaking pipelines, offshore production discharges, and non-point runoff from land-based facilities. Minor "spot" oil spillages occur as illegal discharges from cargo washing. Even if, due to improvements in practices, technology and regulations, a declining trend has been observed during the recent years, oil spills are a common and inevitable occurrence. Therefore, small oil spills—i.e. less than 1,000 barrels—are considered "regular incidents".

According to a worldwide research on oil spillage, about 800 million litres of petroleum enter the ocean each year (Committee on Oil in the Sea 2003). An additional 680 million litres seep into the ocean naturally from the seafloor. Accidental spills from offshore installations account for about 3 percent of the total world oil spills, while spills from oil transportation is now less than 4 percent. According to The Mariner Group,[20] a Norwegian enterprise specialized in oil spill services, during the last decade more than 4.5 billion litres of oil spilled worldwide.

The impact of hydrocarbon spills to marine life last for more than a decade because of the residual oil. Given the hydrocarbon toxicity, even small spills can produce considerable environmental damage. When gas is spilled most of its volume evaporates, while only a small portion dissolves. Currently, only limited data on the impacts of natural gas in the marine environment are available. When oil comes to contact with seawater it undergoes physical, chemical and biological processes including spreading, evaporation, dispersion, emulsification, dissolution, oxidation, sedimentation and biodegradation (Wills 2000). These processes—collectively known as 'weathering'—transform the hydrocarbon into substances with different physical and chemical characteristics.

As soon as oil enters the sea, it quickly spreads over the surface into slicks millimetres thick at a speed depending upon its viscosity (low viscosity more quickly) and the prevailing conditions of temperature, currents and winds. Within the first few days following the spill, crude oil—depending on its heaviness[21]—can lose by evaporation up to 75 percent of its initial volume; however, heavy crude oil loses no more than 10 percent. The oil slick is then dispersed, since waves and turbulence at surface can cause the slick to break up into fragments and droplets of different sizes. The emulsification process[22] follows, forming a viscous compound of oil and water (between 60 and 85 percent of water) called "chocolate mousse". At this stage the volume is increased by three to five times the original volume of the spilled oil. Water-soluble compounds in oil may dissolve into the surrounding water due to dissolution. However, oil contains only small amounts of these compounds, making dissolution one of the less important processes of weathering. The dissolution process takes more time than the other processes, according to hydrodynamic and physicochemical conditions of surface waters. The subsequent oxidation process is promoted by sunlight. Hydrocarbons react chemically with oxygen either breaking down into soluble products, or forming persistent round compounds called tars (tar balls). Since hydrocarbons are usually less dense than seawater, sedimentation occurs because some of the oil (up to 10–30 percent) is absorbed on the suspended material and deposited to the bottom. The end of hydrocarbons in the marine environment is eventually their complete transformation and degradation due to microbial activity. About a hundred known species of bacteria and fungi are able to use hydrocarbon components to sustain their metabolism. The main factors affecting the efficiency of biodegradation are: the structure of hydrocarbon molecules; the level of nutrients (nitrogen and phosphorus) in the water; the temperature; and the level of oxygen.

After all chemical and microbial transformations, oil may aggregate in the form of petroleum lumps, tar balls or pelagic tar. Oil aggregates look like grey, brownish or black sticky lumps that can be found both at sea and on the beach. They have different shapes and usually vary in size from 1 mm to 10 cm. Oil aggregates can last from a month to a year in enclosed seas and up to several years in the open ocean.

Clean-up actions

Clean-up actions are basically ineffective in mitigating the impact of oil spills. On average, they are able to recover 5 to 15 percent of the hydrocarbons (Offshore Oil and Gas Research Group 2004). Regarding clean-up strategies, oil spills may be cleaned up by means of various techniques, even if clean-up efforts have had modest success to date. Spills on the surface can be fenced by floating booms in order to let especially equipped boats to skim the oil away. Another technique consists of spraying chemical control agents over the spill. Dispersants may be used to speed up natural degradation. Shores are often cleaned by hand, which is of course a very labour-intensive method. Occasionally oil-eating bacteria are used (bioremediation) and sometimes hydrocarbons are merely ignited.

GAS HYDRATES

Gas hydrates are a possible future source of energy that can be harvested from the sea. This is perhaps the most significant potential development of deep-sea mineral mining because methane is widely used as a starting material for more complex molecules to use as liquid fuels and lubricants and for chemical compounds. Gas hydrate is a crystalline formation made by gas molecules—usually methane—entrapped in a cage of water molecules. Methane gas is generated primarily by the decay of organic matter from tiny marine plants and animals.

Gas hydrates occur in areas where appropriate conditions of low temperature and high pressure cause water and methane to combine to form a solid. These conditions are found in zones of thermodynamic equilibrium called Hydrate Stability Zones, such as in permafrost arctic regions, in deep-water basins close to continental shelves and in deep ocean basins with a sediment thickness of at least 1,000 m (ISA 2001b).

Experimentally, clean methane in the presence of water temperature of 3° C and a pressure of 2 MPA (Megapascal)[23] forms methane clathrate (methane gas hydrate). A major feature of gas hydrates consists of the high gas concentration in the volume unit of the material, which occurs when temperature and pressure parameters correspond. As temperature increases, pressure must also increase in order to form the hydrates. Methane hydrate in seabed sediments has been mapped at depths below the seafloor[24] ranging from about 100 to 1,100 m. This represents a huge hydrocarbon reservoir placed in the continental margins.[25]

Gas hydrates are widespread over a 90 percent area of the bottom of the oceans. One cubic meter of hydrate yields about 160 cubic meters of methane. Even though estimates of the world's amount of methane in gas hydrates may be overstated—and it is also likely that many hydrate deposits have low concentration and therefore no commercial potential—it is generally believed that the amount of gas in the hydrate reservoirs of the world greatly exceeds the volume of known conventional gas reserves (Collett 2003).

An estimate of the amount of geological reserves of gas hydrates in the world's marine accumulations gives a figure ranging between 300,000 to 90,000,000 trillion m^3. This corresponds to hundreds of times all known world reserves of oil, gas and coal (Kvenvolden 1993). The most conservative estimate of the methane extent at sea figures out an amount of about 10 percent of the known reserves of fossil fuels (UN-ISA 2004). Microbes that dwell in the surface sediments consume some 5–25 percent of the methane deposited. These organisms, called *Archaea*, oxidize methane to provide energy for their life functions and the by-products of this reaction are free hydrogen and carbon dioxide. This process involves a symbiotic relationship with sulphate-reducing bacteria that maintain a low enough partial pressure of hydrogen to keep the methane reaction energetically favourable (Committee on Exploration of the Seas 2003).

Research programs in Japan, India and the United States have been done in order to assess both the availability and the production potential of gas hydrates. The high cost of imported hydrocarbon resources is one of the reasons why government agencies in Japan, India and South Korea have begun to develop hydrate research programs to recover gas from oceanic hydrate. However, productions of gas from hydrate fields are currently limited to Japan and Russia. Gas recovery from hydrates is hindered because hydrates occur as a solid in nature and are commonly widely dispersed in deep marine environments. Proposed methods of gas recovery from hydrates aim to decrease hydrate stability conditions, usually dissociating *in-situ* gas hydrates by heating, depressurizing, or injecting an inhibitor into the reservoir. Among the techniques for natural gas production, the most economically promising method is considered to be the depressurization scheme. Scientific and economic feasibility studies on methane hydrate production indicate that thick vertically stacked deposits limited to small areas are more economically productive than deposits spread in a thin layer across large areas. The National Petroleum Council published in 1992 an economic assessment of gas hydrate production using thermal injection and depressurization. According to this study, hydrates do not appear to be an economical energy resource.

4 Energy from the Oceans

CONVENTIONAL VS RENEWABLE

With a total primary energy supply of more than 11 billion tons of oil equivalent (Btoe) recorded in 2004, the world energy market is a 3 trillion dollar business per year and it is expected to rise significantly over the next decades (EIA 2006). Since energy is essential for the present way of life and any threat to its supply is of major importance to the well being of a country, the energy sector is not a commercially driven market. Regulations and restrictions are established in almost every country to conform the energy market to social and political requirements. To ensure the continuity of energy supply, a country cannot be entirely dependent on just one power source and, consequently, a strategic mix of different energy sources is always present. Since the First World War, the main source of energy has come from oil and gas, which today represents 62 percent of the total world energy consumption (see Figure 4.1). However, conventional forms of energy—in particular fossil fuel—create pollution that damages the environment, its natural resources and the health of its inhabitants. Burning coal for instance—which accounts for 26 percent of total world energy consumption—releases about 1 Kg of CO_2 per kWh[1] in the atmosphere. In 2004 global CO_2 emissions were 26,583 million tons (EIA 2006).

Given these figures, there is increasing international agreement on the risk related to the production of greenhouse gasses and the threatening effect they have on global warming, climate change and other environmental dynamics. The Kyoto Protocol is the last attempt to address this risky situation. The pollution/health issue is one of the main reasons why many efforts to increase the share of more environmentally friendly forms of energy are under way in industrialized countries. However, there are also other reasons, such as the expected scarcity of conventional energy resources in the future and the fact that in many industrialized countries a large part of the energy consumed is imported. In Europe, for instance, 50 percent of the energy is presently imported and the dependence is forecasted to increase in the future to about 70 percent (EU 2000).

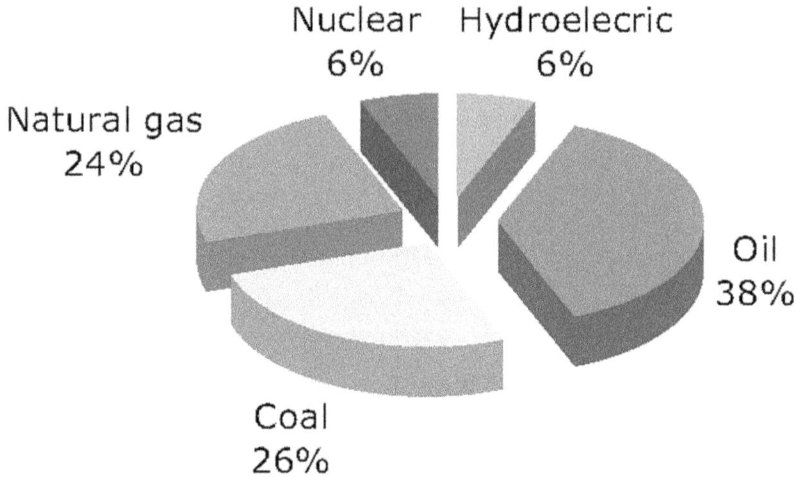

Figure 4.1 World Sources of Energy in percentage, based on British Petroleum Statistical Review of World Energy (June, 2006).

Worldwide efforts to reduce fossil fuel emissions, coupled with the desire of establishing a more autonomous and sustainable supply of energy, have fuelled interest in the development of different renewable energy systems, including marine energy converters. Renewable energy sources are still expansive to harness with the present technological apparatus and, therefore, are not commercially competitive with other well-established conventional forms of energy. However, the costs of producing and using conventional forms of energy do not include the high external cost—i.e. the cost of repairing the environmental and health damage that they provoke. On the other hand, renewable energy plants have very low external costs. Coal, for instance, has been calculated to have 2 to 12 cents US per kWh of external cost, while wind energy is just about 0.12–0.20 cents US per kWh (GWEC 2006). A socially and environmentally oriented comparison should include these costs in the account.

Regarding the marine environment, it is common knowledge that the oceans contain a huge quantity of renewable energy, a reservoir that is theoretically greater than the present world's electrical energy consumption, which is less then 2 TW. Estimates of the full potential opportunity of harnessing ocean energy range between 1 and 10 TW (1 terawatt (TW) = 10^{12} Watt) (EUREC 2005). In the future, this source of energy, thanks to the constant advances of the harnessing technology, will become a financially viable possibility and will contribute to satisfying the world's increasing energy demand.

At present, as noted by Fraenkel (2006)—the technical director of one of the most advanced companies in the sector of marine tidal energy—the large investment required by marine renewable plants could be overcome only if large rewards were expected as in the offshore oil and gas industry. In general, the problem with renewable energy sources is that the technology used for the exploitation must be as inexpensive as possible to be cost-effective, because the resource is diffused and available to anyone. However, if a source of energy provides other advantages, to produce at the lowest cost should not be the main requirement and the preferred technology should be chosen from a social-environmental point of view.

The development of renewable energy projects is similar to other high-tech industries. In high-tech sectors, initial capital cost is high and cost reductions start to occur only as the technology properly matures. In the renewable energy sector, capital cost starts to decrease only when the installed capacity base grows, because the average cost of power generation decreases. There are encouraging examples of successful renewable energy plants, such as the photovoltaic and the wind energy systems, whose technological experience can help to forecast the future of marine renewable energy. European photovoltaic energy data for the period between 1976 and 2002 indicates a steady and progressive decrease in costs (CRES 2000). Wind turbine energy system is already a well-developed form of clean energy in some industrialized countries[2] and its share in the energy market has constantly increased in the last 20 years. Initially based onshore, wind turbine technology also successively began to be developed offshore. At the end of 2005, some 59,000 MW of installed generating capacity were operational in the world (GWEC 2006), and approximately 30 different 2 MW wind turbine devices were commercially available in the market (Orme and Masters 2006). The wind turbine sector has grown at an annual average rate of 35 percent since the late 1990s (EUREC 2005) and in prospect this market is expected to expand its production even faster.

Initial capital cost for onshore wind energy turbine has notably decreased over the past two decades: US figures indicate that from USD 2,800 per installed kW recorded in 1980, capital cost decreased in 1995 to USD 1,000. In the same period, the average cost of electricity from wind-power plants fell from USD 0.51 per kWh to USD 0.47 (IEA 2005). European figures for these energy sources indicate the same trend (CRES 2002).

At the present stage of development, the technologies to exploit renewable forms of energy are still more expansive than conventional ones. As this promising high-tech industry is still in its infancy, the renewable energy sector needs a supportive political and legal framework to help make a breakthrough. Political and financial support to the renewable sector is therefore of utmost importance in order to give the new technologies the possibility to compete on a commercial basis. Many governments are in fact trying to promote a sustainable market for green electricity. Policies include public investment in research and development as well as financial and tax incentives for private sector investment in research and prototypes testing. Since renewable energy facilities are almost always small and geographically spread, another supporting public policy is to facilitate the grid connection. In Europe, after the tenfold increase in investment—from EUR 1.5 billion in 1990 to EUR 15 billion in 2004—the renewable energy sector is more developed than elsewhere (EUREC 2005). The EU has planned to increase its share of renewable energy sources for electricity consumption from 15 percent in 2001 to a target of 21 percent by 2010. This is an ambitious objective, since the rising demand of electricity continuously counteracts progress in the renewable energy share of supply. However, no form of energy—conventional or non conventional—has ever been developed without government support. The result of investing public funds to assist research and development in the renewable energy sector would assist in shaping this market up to the point where private investment will take over and sustain the market itself.

Hydroelectrically powered plants are the oldest source of marine energy. The installed worldwide capacity produces in total a little more than 250 MW. Wind turbine farms can be located offshore and numerous examples are available. However, apart from accessibility problems, they share the same characteristics of the plants located onshore. The energy comes from the wind, which is common both onshore and offshore. Therefore, it is not a peculiar source of marine energy and it is not analyzed here. Even if some harnessing devices have been tested extensively and some pre-commercial technological demonstrations are ongoing around the world, compared to other sources of renewable energy, ocean technology is still at the invention and research stage. However, some types of wave power devices can already experimentally produce electricity in an economic fashion. Financially viable technologies for marine renewable power plants have not yet been developed on permanent basis. Producers are getting closer and closer to commercial exploitation, but the sector needs strong public support to take off. With regards to public support to the sector, the EU started to show interest in ocean energy in 1986, even though it was only in 1993 that a series of conferences on the subject had been sponsored. In 1994 the EU Commission started its own research program, assessing the European marine energy potential. The result was a publication: *The European Wave Energy Atlas* (WERATLAS). Furthermore, in order to coordinate collaboration among European countries in the wave energy sector, the European Thematic Network on Wave Energy was launched in 2000. In figures, research and development funding to marine energy from the EU has been EUR 26 millions in the last 20 years, compared with the tenfold funding of research and development in photovoltaic projects (IEA 2005).

The two main forms of renewable energy coming from the oceans that have potential to be extensively exploited in the recent future are thermal energy and mechanical energy.[3] Thermal energy can be harvested by means of the thermal differential between the warm water at surface and the cold water at depth. Mechanical energy can be transformed into electrical power by means of harvesting the force coming from the motion of the water, including waves, tides and currents. The common originating source of these forms of energy is the sun. As was already stated by Furnas back in 1941, the sun transmits to the Earth in just one minute as much energy as is consumed in the world in one year. The heat from the sun generates both thermal energy that can be harvested from the thermal differential and mechanical energy coming from waves and currents. There are several different systems to harvest these carriers of energy but still none of them can be identified as the leading commercial technology.

The major challenges facing marine energy converters are twofold:

1. To resist the harshness of the marine environment—i.e. the risk of damaging or losing expensive devices; and
2. To convert energy efficiently.

Although the technology for exploiting thermal and mechanical energy from the oceans is still being developed and prototypes abound because of the immaturity of the sector, four main streams of technologies can be identified:

1. Thermal energy can be harvested by means of a system called Ocean Thermal Energy Conversion (OTEC);
2. Mechanical energy can be harvested in three different ways:
   a. By exploiting the wave kinetic energy;
   b. By exploiting the stream of the tides;
   c. By exploiting the flow of the currents.

OCEAN THERMAL ENERGY CONVERSION SYSTEM

The thermal energy generated from the sun that is absorbed and reflected from the oceans can be used to produce electricity. The idea of using the difference in water temperature to produce energy dates back to 1881, when Jacques D'Arsonval, a French engineer, first began to think about the OTEC system. To generate electricity with the OTEC system a temperature difference of at least about 20° C is required. OTEC harvesting technology is, therefore, limited to locations around the tropical regions (between the latitudes 10° N and 10° S) because of the higher temperature of the surrounding surface waters compared to other latitudes. The temperature difference can be used to power a motor that works on the same principle as that of the heat pump.

In order to convert the thermal gradient into electrical energy, the warm water is used to heat and vaporize a liquid called 'working fluid'. The working fluid develops pressure as it evaporates because of the heat. This expanding vapour runs through an air turbine that powers a generator and then is condensed back into liquid form by the cold water brought up from the depths. The cycle is then repeated. A transmission cable connected with the grid carries the electricity onshore.

Within the general framework of the OTEC system three different types of electricity conversion techniques can be found.

1. The closed cycle system
   This kind of converter uses the ocean's warm surface water (close to 30° C in the tropical regions) to vaporize a working fluid with a low boiling point. Evaporation is the thermal separation (or thermal concentration) process used to remove a liquid from a solution, suspension, or emulsion. In the OTEC system (see Figure 4.2), boiling off the liquid through an evaporator performs this process. The steam of the evaporating fluid expands and turns the turbine, which in turns activates a generator to produce electricity. A water pump, through a flexible pipe or riser, brings the 7° C cold deep water (located at around 1,000 m depth) to a condenser placed at the surface. The cold water condenses the steam and the closed cycle begins again. The flow rate of ammonia—which is the most commonly used working fluid—is 31.6 kg/s to produce 1 MW of power across an axial turbine.[4]
2. The open cycle system
   This kind of converter boils the seawater—that in this case is the working fluid—operating at low pressure. The resulting steam passes through a turbine powering a generator in the same way of the closed cycle system. Desalinated ocean water is the by-product of the open cycle process. Theoretically a 2 MW OTEC open cycle plant can produce about 4,000 cubic meters of desalinated water per day.
3. The hybrid system
   This kind of OTEC system combines features of both the closed cycle and the open cycle techniques. Warm seawater enters a vacuum chamber, where it is evaporated into steam exactly like in the open cycle process. The steam, in turn, vaporizes the working fluid in a closed cycle loop. Finally, the vaporized fluid drives an air turbine generating electricity.

The first OTEC system plant was constructed in 1930 at Matanzas Bay, in Cuba's North Coast by the French scientist Claude, a student of D'Arsonval. The prototype produced 22 kW gross of electricity, corresponding to lower than zero net, meaning that the plant consumed more than it produced. In 1956 other French scientists intended to build a 3 MW OTEC plant in the Ivory Coast, but the project was abandoned because it was considered too expensive. In 1979 the Hawaiian Natural Energy Laboratory conducted an experiment with a small OTEC plant onboard a vessel moored 1.5 nm from shore. This was the first successful offshore production of net electrical power from a closed cycle OTEC plant. In 1993,

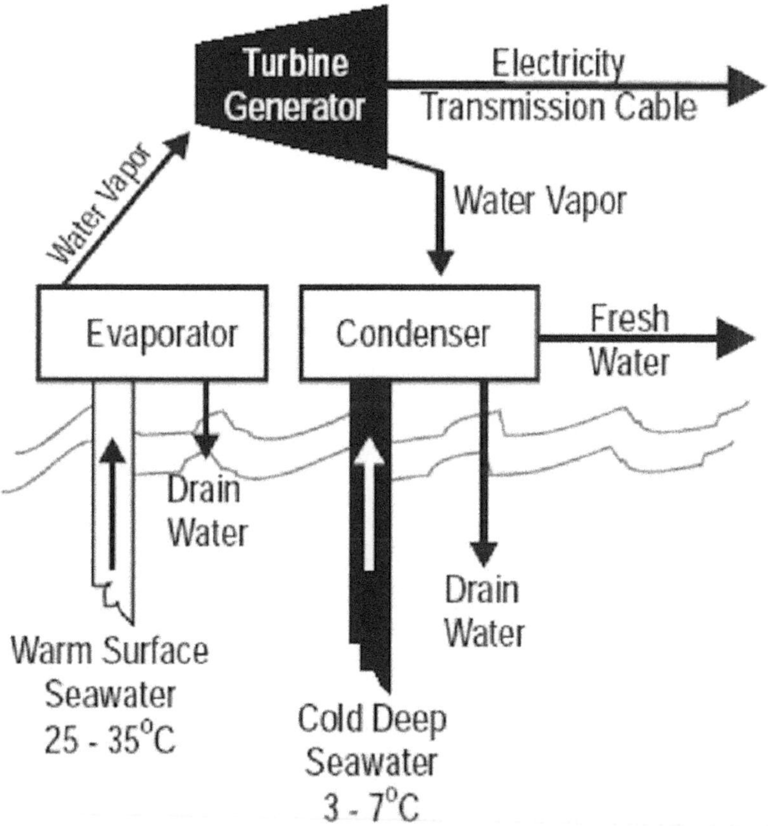

Figure 4.2 OTEC System Operating Scheme. Source: Minerals Management Service, US Department of Interiors, Ocean Energy, (undated).

another experimental plant with an open cycle OTEC system produced 50 kW of net power onboard a ship moored in Hawaii.

The major environmental concerns of OTEC system are twofold: (1) the potential change in the oceanographic properties of seawater due to pumping; and (2) the chemical pollution arising from working fluid leaks and corrosion.

The changes in both water's salinity and temperature could have unforeseen consequences for the local ecology. Researches are still trying to assess the possible impact. Pollution from fluid spills can be overcome using an environmentally friendly working fluid. Generally speaking, a careful site screening is the solution to minimize the environmental impact of offshore or coastal structures. However, in the case of OTEC plants the choice of the site for the plant is limited to few locations.

Three main factors impede the development of OTEC power plants. The first is that the thermal efficiency of OTEC plants—including the power required for pumping the water and losses—is about 2–3 percent. Therefore, present OTEC systems are not very energy-efficient. The principal problem is to bring cold water to the surface. OTEC plants require an expensive, large diameter intake pipe long enough to reach the ocean's depths. Pumping water in a non-consuming way is a massive engineering challenge and the proper technology to efficiently deal with this problem is yet to be developed. Second, the construction of large-scale OTEC power plants involves enormous capital investment and the private sector is often unwilling to make such a financial commitment due to the high risk of the venture. Public intervention is therefore needed if OTEC systems are to be developed. Third, an OTEC plant feasibility is limited to a few hundred sites located in the tropical regions where cold deep-ocean water is close enough to the shoreline.

WAVE KINETIC ENERGY

Waves are moved primarily by winds, which, in turn, are stimulated by solar energy. The different heating that the Earth receives from the sun produces low and high pressure in the atmosphere, which interaction produces winds. Winds transfer some of the energy to the water through friction between air and water molecules. The size of the waves depends on the wind speed, the length of time and the distance—or fetch—the wind blows. There are great variations in wave power level but in principles the power available in a wave is proportional to the square of the wave height (amplitude) and to the wave period, or period of motion. A commonly used yardstick to rate waves' potential is the energy flux, which refers to the energy in kilowatt transported in each meter of wave crest length. Long-period, large-amplitude waves have an energy flux of 40–50 kW/m (CRES 2002). The global wave power potential, which is the greatest among marine renewable, has been estimated between 8,000 and 80,000 TWh per year (IEA 2003), corresponding to 1–10 TW. The World Energy Council estimates the wave potential at 2 TW (Thorpe 1999).

Wave power varies seasonally and the wave resource at the development site is, of course, a critical factor in assessing the efficiency of the power plant. Wave energy is at its strongest power in ocean depths of at least 50 m, before the wave loses energy because of the friction with a shallower seabed. The best locations for exploiting wave power are found at temperate latitudes between 30° and 60°. Waves considered useful for energy production vary from 40 to over 300 meters in length (EPRI 2004) and the best sites for wave-power plants must have more than 30 kW/m^2 of wave power. More powerful waves (up to 70 kW/m and more) would destroy any device (see Figure 4.3).

Wave Energy Converters (WECs) transform the kinetic energy of the moving water into electric power through a generator. The power input is

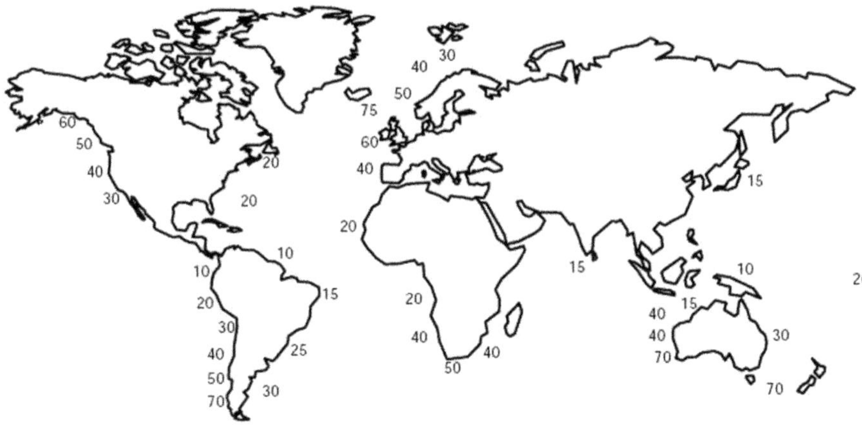

Figure 4.3 World Wave Energy Source. Wave Power in kW/m. Source: IEA-OES (2003).

the low-speed, high-force wave motion that must be converted to the high-speed, low-force power output of the electric generator. This mismatch can be overcome in different ways. Current technology is able to achieve power conversion by means of hydraulic, hydrodynamic and pneumatic devices. Even though there has been extensive research and development of WEC technology, it only began in the 1970s as a reaction to the oil crises and the subsequent rally of energy market prices. However, in France, way back in 1799, Girard & Son patented a technique to convert the ocean's wave energy into a useful form of energy. In theory, as water is 832 times denser than air, it should be possible to get the same amount of energy from the water motion with a device much smaller and cheaper than a wind driven turbine. Although similar to wind turbines, underwater turbines are in fact much smaller and have a slower motion. On average, a water turbine has a dimension of about 9 m against the 90 m of a wind turbine and turns at about 30 rpm (revolutions per minute) against the 60 rpm of a wind turbine. In 1994 the EU Commission started a research program on WECs technology, including studies, financing projects with technical feasibility and the formation of the European Thematic Network on Wave Energy in order to coordinate collaborations among European countries. WECs are expected to provide in the near future an efficient, reliable, cost-effective and environmentally friendly source of electrical power. Some European governments, in particular the UK and the Scandinavian countries, have programs to develop this renewable energy sector by means of financing scientific research and experiment of prototypes.

Although there has been a recent explosion of different devices and great expectations all over the world, the wave energy extraction is still a nascent technology and the route to bring the product to market still looks long. Over 1,000 wave energy conversion techniques are patented in Japan,

North America and Europe, testifying the immaturity of the technology in this sector. However, there have been significant reductions in the estimated costs of the devices and some WECs have already the potential to provide cheap electricity for small islands and remote coastal communities (Pontes and Falcao 2001). Regarding investment cost, WECs power-rated between 750 and 1,000 kW of installed capacity have a market price of around USD 3–4 million. Deployment, mooring system for offshore devices and electrical interconnection costs are excluded (EPRI 2004). With regards to operating costs, it has been said that to generate cost-competitive WECs the technology should be capable to reduce production costs from the current USD 0.75–0.8 per kWh, to the wind-driven turbines' level—which is around USD 0.45–0.5 per kWh (Thorpe 2000). To compare the production costs of renewable energy systems with conventional ones, it is worth reminding that production cost for large-scale coal burning power plants is about USD 0.25–0.3 per kWh, while combined cycle natural gas-turbine technology production cost is about USD 0.3–0.5 per kWh (IEA 2003).

With WEC technology there are three steps in the wave energy conversion process. The first is the primary conversion, in which the energy from the motion of the waves is transmitted to an absorber structure. Next is the secondary conversion, in which the absorber structure transmits the energy to a mechanical generator. The third is the conversion from the energy produced by the mechanical generator to an electric generator (IPRI 2004). Each WEC is composed of several sub-systems, or components, all of which must be suitable to survive the weathering effects caused by the salinity and the severe environmental conditions of the ocean.

1. The structure
   The WECs' structure contains and protects all the mechanical and electrical parts from which the device is composed. WECs often have a steel structure that is sandblasted and then painted with a protective coating to reduce biofouling. However, other structural materials common to offshore construction and shipbuilding such as concrete and fibreglass reinforced with polyester are used as well. Offshore steel structures have proved to have a longer life than concrete ones.
2. The absorber structure
   Also called activated body or prime mover, the absorber structure is the device that reacts to the motion of the waves and is responsible for the absorption of the kinetic energy. There are different ways to convert wave energy into electricity, however, all of them rely on the prime mover, which is the real 'engine' of the device. For instance, the activated body can be a device such as: (a) the capture chamber in Oscillating Water Column (OWC) converters; (b) a power buoy placed at the surface directly connected to a generator on the seabed (direct driven generator);[5] (c) two or more buoys connected with joints in which hydraulic oil is pumped in response to the motion; or

(d) a flap like in the pendulor. The activated body utilizes the motion of the waves to activate the power take-off system.

3. The power take-off system (PTO)
The PTO system is the main generator that converts the slow oscillating movements of the activated body into electricity. It includes mechanical and electrical components such as pumps, hydraulic motors, turbines, gears and the electric generator. The PTO system is designed according to the type of conversion the prime mover performs. Therefore, there are different kinds of PTO systems. If the prime mover converts the wave motion into airflow pressure—such as in OWC converters—the PTO system must be an air turbine. If the prime mover channels the wave motion into water flow pressure, the PTO system must be a water turbine. When the prime mover converts the wave movements into relative motion among the different parts of connected buoys, the PTO system can be either a hydraulic pump or a mechanical transmission such as a rotor.
In the electric generator, these mechanical movements spin coils of wire inside ring-shaped magnets to produce current. All electric generators utilize a translator that activates a magnetic field to produce the electric current. The electric generator of the PTO system can be rotating or linear, depending on the movement of the activated body. A conventional generator has a rotating rotor, while in a linear generator the motion of the inside translator is up and down. Electric generators can be divided into two groups: Alternating current (AC); and direct current (DC).
4. The electrical connections
There are two main connection systems in WECs: the connection within the individual units that compose the system and the connection of the system with the grid. Several WEC units may be interconnected in a farm in order to increase the total output and reduce the number of transmission cables to the grid. Each unit is connected through a flexible riser cable to a common junction box located on the seafloor. The junction box is connected to the grid. The grid interconnection is the infrastructure required to connect the wave farm to a nearby land-based grid interconnection point. Wave generators convert mechanical energy into electrical energy. However, the generator works with a power source—i.e. the prime mover—that supplies irregular mechanical power because of the random nature of the waves. Therefore, if the PTO is either a wind turbine which runs at variable rotational speed, or if the generator is direct driven, such configurations produce a varying alternative voltage—where both amplitude and frequency change constantly. In this case, indirect grid connection is required. This means that the variable frequency of the WEC AC generator must be first converted into DC using a frequency converter (rectifier) and then converted again (with an inverter) into

an AC that has exactly the same frequency and transmission voltage as the electrical grid. The electrical loss due to interconnection is similar to other power plants, about 2 percent of the production.

5. A mooring component
   This is, of course, necessary only in offshore power plants and it is required for holding the device in place. WEC mooring systems almost always tend to conform to Lloyds Register Rules for floating offshore installations at fixed locations. This usually consists of a number of catenary (chains) mooring legs connected to seabed-located clump weights (EPRI 2004).

As already mentioned several times, marine equipment must confront a heavy weathering environment, therefore, operation and maintenance activities have great importance because of the high costs involved. Given the general difficulty of accessing offshore facilities even on the surface and the major complexities of assessing submersed components, ease of operation and maintenance is one of the main concerns in offshore wave energy converters. Generally, operation and maintenance activities of offshore devices are expensive and can be carried out only when weather permits. In particular, the cost to repair sub-sea components is extremely high, therefore, bottom-standing systems are the most expensive to maintain. Simplicity of the offshore device is an asset because complexity almost always increases operation and maintenance activities.

An indicator of the complexity is the number of components forming the system. The remote supervisory and control capability of the device is also a plus, as it minimizes the physical intervention requirements. Some offshore WECs feature extensive remote diagnostic capability able to identify problems and failures from the onshore control station. The large dimension of some devices might permit access by helicopter, thus improving its accessibility, especially during unfavourable weather condition. A number of wave energy devices employ direct induction generators (that do not need to store energy in accumulators and convert it) thus significantly reducing operation and maintenance requirements (EPRI 2004).

It is possible to speculate that average maintenance costs of WECs would be similar to the wind energy converters of the first generation, plus an additional cost related to the offshore location and the presence of submerged components. Old wind turbines had an annual maintenance cost of about 3 percent of the capital investment. WECs would probably have an average annual maintenance cost of about 5–6 percent.

Ocean wave energy technology is based on different systems that exploit the motion of the waves in numerous fashions. The main common feature of WECs is to convert the low-frequency energy coming from the waves to the 50–60 Hertz frequency compatible to the electric grid. WECs can be classified according to their size and orientation in relation to the wave direction. Devices comparatively small to the wavelength are called 'point absorbers'.

Floating structures of approximately the same dimension of the wavelength are called: (A) 'attenuators', if aligned in the direction of the waves; and (B) 'terminators', if perpendicular to the direction of the waves.

Another way of classifying WECs is to distinguish them according to the location in which they are based. WECs can be based in two different locations: (1) on the shoreline; (2) near-shore and offshore. Each location hosts a different kind of converter. However, this distinction is not rigid because some devices may be based both onshore and offshore. Shoreline systems have the advantage of ease-of-access for both installation and maintenance, while offshore systems benefit from the more powerful waves, as wave energy rapidly decrease in shallow water.

A. Shoreline systems

There are three main types of WECs located in the shoreline that are presently developed as experimental plants or are still under development.

1. Convergent Channel System or Tapered Channel System (TAPCHAN)
   In this type of converters, the waves are funnelled by a shore-based structure into an elevated artificial lake or reservoir (see Figure 4.4). The waves are amplified in a narrowing collector until the wave-crests

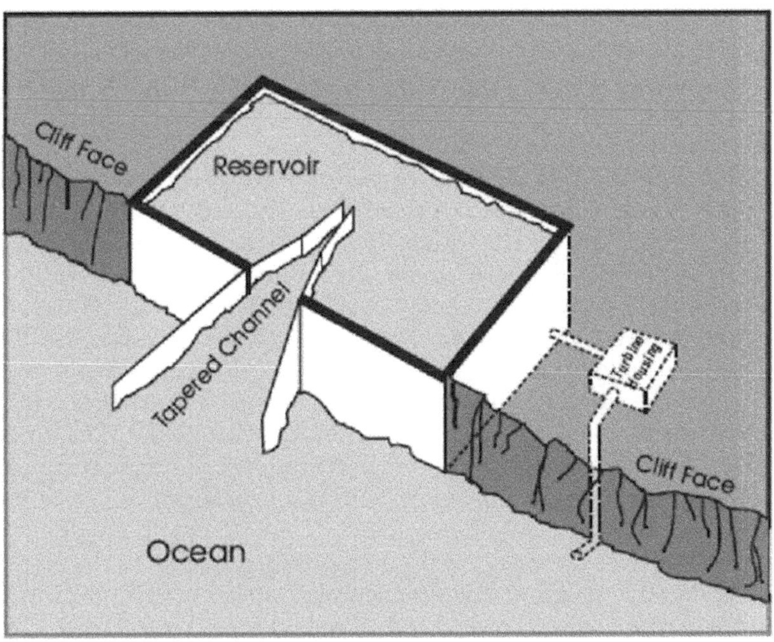

Figure 4.4 The TAPCHAN System. Source: Minerals Management Service, US Department of Interiors, Ocean Energy, (undated).

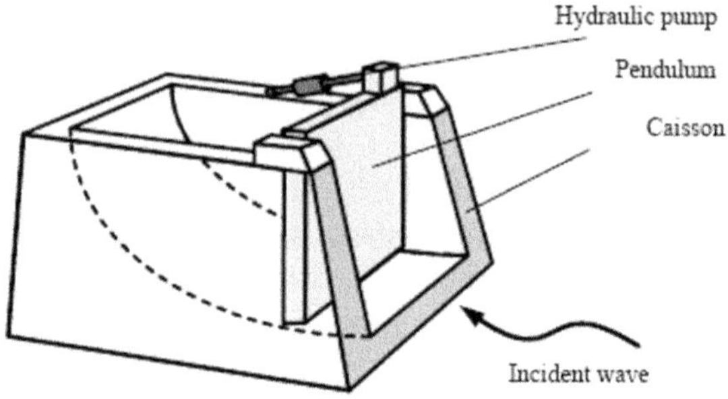

Figure 4.5 The Pendulor. Source: Minerals Management Service, US Department of Interiors, Ocean Energy, (undated).

spill over the wall of the reservoir. The water flowing out of the reservoir is then used to produce electricity with standard hydropower technology (water turbines).

2. Floating system or Pendulor
   This system generates electricity from the bobbing action of a floating object that drives a hydraulic pump (see Figure 4.5). The floating object can be fixed on the seabed or mounted to a moored floating platform. The device consists of a rectangular box (or caisson), which is open to the sea at one end. A pendulum flap is hinged over this opening so that the action of the waves causes it to swing back and forth. The motion of the pendulum is used to power a hydraulic pump that in turn powers a generator. In China more than 600 units of this kind have been deployed along the coast in the past 13 years.
3. Oscillating Water Column (OWC) system
   Shoreline OWC is the most employed wave energy technology, with several large-scale prototypes installed and successfully operating in the Azores (Portugal), China, India, Japan, Sweden and the UK. A former Japanese Naval Commander invented this system in 1947. The technology used for the energy conversion is pneumatic power, therefore, an OWC extracts the energy through an air turbine (see Figure 4.6). A hollow pipe, or collector, is placed with its lower opening below the surface. The water enters the pipe and when the incident waves move inside the water column, this 'oscillates'. The air inside the chamber and above the water column is compressed and turns a Wells turbine,[6] which, in turn, powers a generator.

In OWC devices the energy is produced in three stages. In the first stage, the wave flux is first converted into pneumatic power in the collector. In the second stage, pneumatic power is converted into mechanical power. An air turbine converts the air pressure from the oscillating water column chamber into rotary motion. In the third stage, the mechanical power is con-

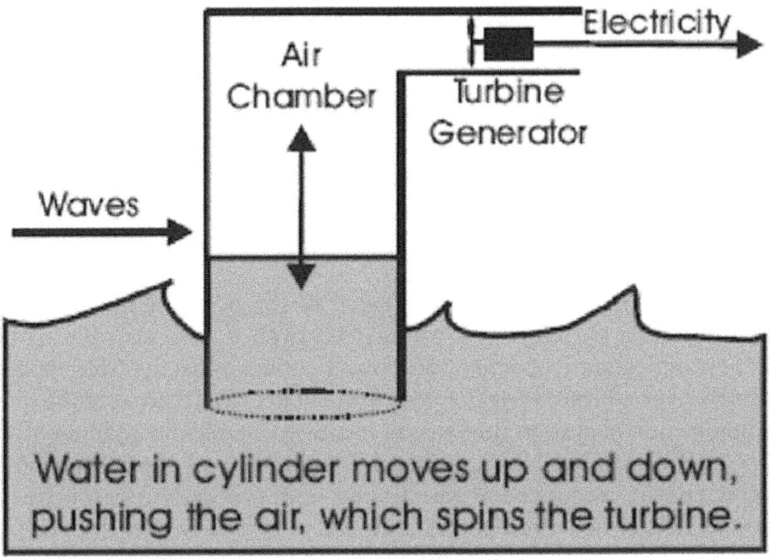

Figure 4.6 Oscillating Water Column. Source: Minerals Management Service, US Department of Interiors, Ocean Energy, (undated).

verted into electric power through a generator. The amount of wave energy captured in the compression of the air in the collector out of the total incident wave energy flux is called capture efficiency. The proportion of pneumatic power that is converted to mechanical power in the turbine is called mechanical efficiency, while the proportion of power that is converted from mechanical power to electrical power is called electrical efficiency.

There are inherent losses of energy during each conversion process. The product of capture efficiency, mechanical efficiency and electrical efficiency is called wave-to-wire efficiency and includes all energy losses. Both estimated and achieved wave-to-wire efficiencies vary broadly from a device to another. Measured efficiencies of near-shore OWC devices have been less than 10 percent, whereas estimated efficiencies exceed 30 percent (The Carbon Trust 2005).

The structure must sustain an extreme and varied loading regime. The most important requirements of an OWC structure are: (A) to create a chamber of air suitable for an efficient compression by wave action; and (B) to provide protection to turbines and generator.

The water column must be sufficiently long that water does not rise high enough to flood the rest of the device. The structure of the OWC converter must be adapted according to each location characteristics because the collector chamber's efficiency depends on the wave climate, which in turn depends on the water depth, on seabed friction losses and other local shoaling effects. A shallow seabed is more likely to create more energy loss than a steep sloped seabed. Therefore, the OWC chamber should be sized to suit site-specific wave conditions.

All OWC devices have a large collector chamber with sub-divisions in the chamber to enhance the energy capture efficiency. Basically, the efficiency of the energy conversion process in OWC devices depends on the matching of the wave frequencies to that of the resonant frequency of the water column. Since variations in wave conditions determine variations in air velocity, an accurate resource assessment should be made prior to the design of the device.

The water column should be adjusted to the peak energy frequency of the waves under different sea conditions. Even if it is possible to obtain valve arrangements to rectify the flow, turbines must be able to operate efficiently over a broad range of inlet flows. This would prevent the need for a complex control system that would increase complexity reducing the reliability of the device. Changing the physical profile of the column in order to maintain the frequency at the desired level can serve this purpose. For instance, an inclined chamber column offers an easier path for the incoming waves and results in less turbulence and lower energy loss.

An example of a sophisticated capture chamber is the OWC converter OreCons MRC1000, which is a freely floating device developed by the British company WaveGen.[7] It is composed of multiple resonant vertical capture chambers of various lengths that oscillate with different frequencies according to the length of each chamber. The multiplicity of the resonant chamber allows this converter to synchronize over a wide range of different wave frequencies. Regarding damage prevention, a valve closes when the waves become too rough providing protection to both turbine and generator. An aspect that increases the lifespan of the OWC system is that none of the moving parts are in contact with the water.

Even if OWC converters can be located near-shore, they are usually based in a suitable cliff. In order to serve a dual purpose, the structure of the device may be incorporated into a breakwater or other offshore structures, such as wind turbine foundations. Examples of the integration of wave energy devices with other marine structures can be found in Asia. For example, the 60 kW OWC project at Sakata Port, Japan, that has been operational since 1989 was built as part of the harbour infrastructure, and integrated in the harbour breakwater wall. Another example of integration with other functions is the "Mighty Whale". With its 50 m in length, 30 m in width and 12 m depth, the Mighty Whale is the world's largest floating OWC prototype, developed by the Japan Marine Science and Technology Center (JAMSTEC). Put into use in 1998 off Gokasho Bay, Japan, it has a rated power capacity of 110 kW and also serves as a wave breaker.

One of the main problems with shoreline-based OWC system is their construction, because these converters must be built on rocky shores exposed to wind and waves. So far, OWC implemented projects have demonstrated that the site preparation and civil construction cost is about 70 percent of the completed project (The Carbon Trust 2005). For this reason, OWC converters based on platforms near-shore are under development.

The WEC developed by Offshore Wave Energy Ltd is an example of the many pneumatic conversion technology devices that have recently been developed offshore. This device was designed to exploit the horizontal component of the wave motion by means of a large 200 m concrete platform moored offshore. Mathematical models of the full-scale performance for a platform located in UK waters resulted in a forecasted output of 4 MW. The project for the implementation of this converter has already reached the demonstration phase, which involves the construction, deployment and feasibility of a three-quarter-scale seagoing demonstrator. The next expected step is the realization of a full-scale commercial platform at the approximate cost of USD 35 million (Kemp and O'Nians 2006). Many small-scale navigational buoys employing the OWC system have operated successfully over many years and proven to be low-maintenance facilities.

B. Near-shore and offshore systems

In offshore and near-shore based WECs, the main feature is an oscillating body that either floats or is submerged. These devices aim to exploit the more powerful waves. The average water depth to locate offshore WECs is about 30 to 50 m or even more, according to the specific characteristics of each device. The two main problems of offshore devices are the anchoring and the connection to take the electricity to land. The distance from the shore significantly increases transmission-cabling cost.

Offshore WECs can be divided into three main broad categories: (1) bottom-standing, (2) freely floating, (3) or semi-submersed.

In bottom-standing devices, the foundation is either incorporated in the coastline, in a breakwater structure, or directly in the seabed. When the converter is fixed to the seabed, this must be prepared with a concrete platform suitable for the required connections.

Freely floating and semi-submersed offshore WECs must be steadily anchored to prevent the device to brake loose of its mooring and drift away when strong forces apply during stormy weather. Some floating WECs are equipped with a ballast to be filled with water, so that in stormy weather the device can submerse to avoid the more severe conditions at the surface. Once the meteorological conditions are suitable, the ballast is emptied and the converter comes back to the surface. Loading and downloading of the ballast is remotely controlled. Another kind of offshore converter is the so-called overtopping device. The overtopping device is a container open on two sides, into which the waves spill water. The entering waves make a water turbine (similar to a hydropower turbine) move and then are channelled back into the ocean.

Apart from navigation buoys—which have been powered with wave energy since the late 1960s—there are many different forms of offshore floating and semi-submersed converters devised by different manufacturers and presently at different stages of development. These include devices such as the University of Edinburgh's Salter's duck, the first offshore WEC to be developed, the Archimedes Wave Swing, developed in the Netherlands by Teamwork Technology, the Pelamis, the Wave Dragon, the Waveplane, the Wavebox, the gyroscope, the floating wave power vessel etc.[8] Although several full-scale prototypes of shoreline and near-shore devices have been tested since the mid-1980s, presently there are only few full-scale pre-production prototypes operating at sea. No full-scale grid-connected offshore wave power plant has been extensively tested so far. The first near-shore (1 km distant at 15 m depth) demonstration power plant to use wave power was OSPREY (Ocean Swell Powered Renewable Energy), which began operating experimentally in northern Scotland at the beginning of August 1995 with a 2 MW capacity.[9] In 1995 a total of 685 kW of grid-connected wave-generating capacity systems were operating worldwide. In 2002 the world wave energy installed capacity grew to 2 MW (CRES 2002). This capacity came from eight demonstration plants ranging in size from 20 kW to 350 kW. In 2000 a 500 kW power rated plant called LIMPET 500[10] (Land Installed Marine Pneumatic Energy Transformer) that works with the OWC system was installed on Scotland's west coast Isle of Islay.

Always in Scotland, the snake-like floating device called Pelamis—which in Greek means sea snake—has been recently connected to the UK electric grid, generating a peak output of 750 kW. The Pelamis device is an offshore semi-submerged WEC manufactured by the British company Ocean Power Delivery.[11] It features an articulated pipe-like structure composed of three cylindrical sections linked by hinged joints. Each section is 40 m long and 3.5 m in diameters. The wave-induced motion of these joints is resisted by hydraulic rams, which pump high-pressure oil through hydraulic motors, which in turn drive electric generators to produce electricity. The Pelamis project has been funded USD 5 million by the UK government and has attracted private investments from an international consortium of venture capital firms of more than USD 12 million (EPRI 2004).

Another operational plant is the European Pilot Plant built into a small natural gully near Porto Cachorro on the Pico Island in the Azores, developed by the Istituto Superior Tecnico of Lisbon, Portugal and sponsored by the EU Commission. Construction works began in 1996 and the plant was commissioned at the end of 1999. This 400 kW OWC plant was designed as a full scale testing facility, however, it is also used to supply, on a permanent basis, part of the island's energy demand.

The environmental impact of WECs is very limited, especially if compared with conventional energy harvesting technologies—a wave energy converter releases about 20 g of CO_2 per kWh (Duckers 2006). The main environmental issues with offshore wave energy conversion facilities are associated with destruction and disturbance of the resident marine life and with degradation of scenic ocean front views. Another possible disturbance is the threat to navigation. Wave energy devices have typically a low profile above the surface and are therefore both unnoticeable by direct sighting and undetectable by RADAR. However, signalling buoys can be effectively placed around the facility area to prevent collisions. Some devices use biodegradable hydraulic fluids in order to minimize environmental impact in case of a leakage.

TIDES AND CURRENTS

Tides are driven both by the gravitational pulls of the moon and the sun, and by the Earth's rotation. They can be predicted decades in advance. The moon, which is located approximately 386,240 km distant from the Earth,[12] exerts a greater influence on tides than the sun does, because the sun—even if it is much bigger—is 150 million kilometres distant from the Earth. Therefore, although the strength of the sun's gravity is 179 times than that of the moon, the moon is responsible for about 56 percent of the Earth's tidal phenomena.

Depending on the season and other local factors, a tidal phenomenon exerts a force capable of shifting near shore water level up to 12 m. The tides generate movements of water into and out of bays and estuaries, creating a vertical water movement. As has been known for centuries, the moving water from tides can be harvested with some device and transformed into mechanical energy. Apart from other uses of minor interest, the energy from tidal streams can be transformed from kinetic to mechanical energy and finally into electric energy through hydropower plants. However, to generate energy from tides in an economical way, large tidal differences—at least of about 7 m—are required. This fact reduces the commercial possibility of exploiting tidal energy to few geographical sites in the world, such as some locations on the Atlantic coasts of the UK, France and eastern Canada, the Pacific coast of Russia, Korea, China, Mexico, Chile, along the Patagonian coast of Argentina, Western Australia and Western India. Historically tidal projects had used large-scale barrage systems to block rivers' estuaries. Tide mills were common in Western Europe (France, Spain and Great Britain) during the Middle Ages, but some date back as far as 700 BC. The total world potential for tidal power has been estimated at 64,000 MW (EUREC 2005).

Apart from the traditional barrage system, energy can also be harvested directly from marine currents. Marine currents are a form of kinetic energy originating from two different phenomena. Global thermal movements create marine currents such as the Gulf Stream and some other large-scale currents, which flow is unidirectional and independent from gravitational forces. However, the same gravitational forces that create vertical water motion (tides) create also horizontal water motions resulting in small-scale bidirectional currents. This kind of current is more intense where coastal morphology creates natural narrow passages—i.e. straits and channels—or where the bathymetry changes abruptly, as the velocity of the current increases with the friction of shallow bottoms.

There are only a few areas with enough average tidal current velocity to be cost-effectively harvested. Assessments of economic viability of tidal stream projects indicate that the velocity of the current should be more than 4–5 knots, corresponding to 2–2.5 m/s (Triton 2002). Therefore, to convert energy from the tides into electric power, two main methods can be used: (1) tidal current turbines; and (2) tidal stream generators. In both methods the conversion into electricity involves the use of mechanical devices such as water turbines.

The tidal current system is also called tidal barrage. The technology used with this system is similar to the technology employed in traditional river hydroelectric power plants, the main difference being a lower head[13] dam, which implies an axial-flow turbine with a low speed and with bigger dimensions. In essence the system is quite simple. A dam is constructed across an inlet (usually a river's estuary) to obstruct the falling tide. Water entrapped into the reservoir at high tide is then channelled back from the higher level of the reservoir to the lower level of the sea passing through water turbines that in turn activate electric generators. The underwater turbines can be mounted on the seabed or suspended from floating platforms.

The operating cycle of a tidal plant consists of four phases:

1. During the rising tide—or flood tide—the water passes through the gates of the sluice in order to fill the basin or reservoir;
2. When the flood tide is at its apex the sluice's gates are closed. The receding tide—or ebb tide—creates a suitable head between the basin and the sea;
3. Water flows from the basin on the ebb tide passing through the turbines;
4. Finally, once the head is reduced to the minimum operating point, the gates are closed again until the tide rises sufficiently to repeat the cycle.

Presently there are only four tidal plants currently operating in the world. The biggest one is the plant La Rance Station in France, built in 1966, which produces 240 MW employing 24 bulb-type Kaplan turbines of 10 MW each. The second largest tidal station is the Annapolis station in Nova Scotia, Canada, built in 1984, which produces 18 MW. The other two tidal plants are the 3.4 MW Jianxia Station built in China in 1986 and the 400 kW experimental plant at Kislaya Guna, on the Barents Sea in Russia, built in 1968.

A tidal energy plant based on a barrage system is a capital-intensive investment with a long construction time. However, plant lifetime can be very long, more than 100 years for the barrage structure and about 40 years for the rest of the equipment. The technologies used to exploit tidal energy are usually made up of conventional engineering systems and more efficient converters should be developed in order to enhance their reliability and to reduce the operating cost. The main environmental disadvantage of the tidal power system is the effect a tidal station can have on plants and animals living within the estuary and the scenic impact of the dam. Since few tidal barrages have been built up to now, very little is known with certainty about the overall impact of tidal power plants on local biological communities. Emission of CO_2 for tidal barrage systems is about 12 g per kWh (Duckers 2006), therefore, negligible if compared with conventional power plants. Nevertheless, given the high cost of building a dam, the related environmental problems and the relatively high unit cost per installed kW, barrage systems are not commercially competitive with conventional fossil-fuel alternatives. They are inefficient, even if compared with WECs, and are no longer considered feasible. As a result, there is presently not much interest in future development of tidal barrage systems and the focus on tidal energy has shifted from the barrage system toward harvesting tidal-generated coastal currents and non-tidal ocean currents.

The tidal stream generator system is also called a tidal stream turbine; it is basically a submersed watermill. The underwater currents created by the rise and fall of the tides rotate a slow-moving underwater turbine, or rotor, which in turn revolves a generator producing electric power. Thus the technology for the exploitation of marine currents is basically a turbine rotor set normally to the direction of the flow and the power produced is a function of both the turbine rotor diameter and the velocity of the current. There are two main types of water turbines to harvest tidal streams: (1) the vertical axis turbine system, also known as axial flow turbine; and (2) the horizontal axis turbine system—or cross flow turbine. The main difference between the two systems is the orientation of the shaft of the rotor relative to the seabed. The tidal stream plant should be located near shore, where this kind of current forms, usually in water depths of up to 50 m. The turbine rotor can be mounted on the seabed or suspended from a floating platform.

First proposed during the 1970s, this technology is presently under development and prototypes are numerous. Marine Current Turbine Company had a tidal turbine prototype plant of 300 kW rated power, called Seaflow, which worked for two years not grid connected in Lynmouth, UK. Seaflow successfully tested the basic concept of an axial flow tidal turbine and a new prototype called Seagen—power-rated 1.2 MW—is now intended to prove the commercial potential of this system in Strangford Narrows, Northen Ireland (Fraenkel 2006). However, the first marine current turbine plant to be permanently connected to an electricity grid is located in the Strait of Messina, Italy and was realized by the Italian company Ponte di Archimede SpA. The plant employs Kobold turbines mounted on vertical axis. The Kobold turbine is a hydraulic turbine with self-moving blades that always rotate in the same direction, no matter what is the direction of the current.[14] This small plant has a maximum power output of 25 kW. The hosting platform is moored to the seabed in a water depth of about 30 m and 150–200 m away from the shore, where the current maximum speed is around 2 m/s. It has been operational since June 2001 and the electricity produced was first used onboard for experimental purposes. In January 2005 an inverter was installed to meet frequency, voltage and phase requirements of the local grid and the plant was finally connected on a permanent basis (Eriksson et al. 2006).

Forecasts about the energy cost of tidal stream power plants have been predicted by a number of consulting companies. Assuming a capacity factor—which is the ratio between the mean generated power and the rated power—of 20 percent and a current speed of 3.5 m/s, Triton (2002) gives figures of about USD 10 cents and USD 22 cents, respectively, for a large (800 MW) and a small (43 MW) power plant. However, future tidal current energy cost is expected to decrease. PB Power's (2006) figures for tidal stream energy capital cost are between USD 3,500–5,500 per installed kW (according to the installed capacity) for the initial period, reduced at USD 1,500–2,000 after the technology has matured. Annual operational and maintenance cost are forecasted respectively at USD cents 3.5 and 1.75 per kWh.

There are some advantages in the exploitation of tidal energy when compared to wave energy, such as its predictability, the steadiness of the flow and the non-extreme environmental conditions of the plants' location. The more favourable conditions, in turn, influence the cost—as the device is not expected to undergo intense atmospheric fluctuations and do not need over-dimensioned resistance.

Within tidal energy systems, tidal current generators are considered to be more efficient than tidal barrage. Tidal stream systems do not require the construction of a dam and therefore are cheaper and more environmentally friendly. Moreover, they are also more efficient because the production of electricity is performed with both ebbing and surging tides. Besides, the technology is modular and therefore does not require large engineering works.

Another new promising area of development in the marine renewable energy sector is linked to the harvesting of large-scale ocean currents. Due to their periodic nature, low visibility and the adaptability of similar well-developed technology, ocean currents exploitation has a great potential. Marine non-tidal currents originate by a complex process related to the absorption of solar radiation in the ocean and atmosphere. The fastest oceanic currents are found on the eastern coasts of continents, such as the Gulf Stream in the Atlantic, the Kuroshio off Japan and the Agulhas-Somali system off the African east coast. Even though oceanic current extraction is convenient only in few specific locations, the total power of ocean currents has been estimated to be about 5 TW (Pontes and Falcao 2001).

There are some common features of marine energy systems. Compared to other renewable forms of energy, such as solar and wind power, which rely much more on atmospheric conditions to work, marine energy is steadier and therefore more reliable. Renewable energy systems implemented to exploit the ocean's energy are characterized by high capital costs and low running costs. However, the cost of electricity is in general strictly related to the initial capital expenditure for the harnessing equipment. Since the fixed cost to install devices at sea is high, economic assessments have showed that to be commercially viable and deliver reasonable returns, offshore renewable systems need to be power-rated at least one MW or more. Components reliability is of major importance, as downtimes caused by components' failure and consequent unplanned maintenance can significantly increase cost of the energy produced. Operation and maintenance cost, which includes insurance, labour and spare parts, are especially high in offshore locations. However, the offshore oil and gas industry has already developed reliable mooring systems, more resilient and corrosion-resistant equipments and enhanced electric underwater cables. These techniques can be usefully employed in the renewable marine energy sector. Proximity of the developing location to electric centres or infrastructure is an asset because long-distance connections to the grid are costly. Regarding the environmental effects of oceans' energy exploitation, although marine power premises are much more environmentally friendly than other forms of conventional energy, it is clear that various natural processes might be affected, especially by large-scale projects.

Natural processes that can be affected include sediment transport and the functioning of coastal biological communities, including marine mammals and seabirds. Some plants may provoke high-frequency (from 70 up to 90 Db) noise pollution and may also represent a visual intrusion on the seascape, negatively affecting the economy of tourist areas. On the other hand, power plants may constitute artificial reefs for colonization from algae and invertebrates, as well as nesting sites for birds, thus having a beneficial impact on marine biota. The possible source of pollution from marine renewable energy plants depends, of course, on the conversion process used to harvest the energy, but, in general, emissions are close to nil. All in all, marine power is a source of renewable energy that must be used extensively in the near future, as soon as the harnessing technology will permit the creation of financially viable plants.

5 Underwater Mineral Deposits and Placers

The ocean waters—both the continental margins and the deep ocean basins—contain unknown quantities of mineral deposits, most of which are the result of geological processes that require at least thousands and sometimes even millions of years to complete. Hence, they can be classified as non-renewable natural resources.

The oceans contain many different minerals, which genesis is diverse. Some mineral deposits derive from mechanical and chemical erosion of land and therefore are said to have *terrigenous* origin. Metallic sediments wearing down from rocks on dry land are carried into the oceans by rivers in particles or dissolved form and then deposited in continental margins by water movements such as waves, tides and currents. This process occurs in the placer formation of gemstones, heavy metallic minerals, beach aggregations of sand and gravel and phosphorite deposits.[1]

Commercially valuable heavy metals are usually found at continental margins in near-shore areas and include barium, chromium, copper, gold, iron, lead, rare earth elements,[2] tin, titanium, thorium, tungsten and zirconium. Minerals such as sand and gravel, corals, pearls, methane hydrate, and oil and gas have instead a *biogenic* origin, since they derive from sedimentation of organic matter over millions of years.[3]

Some other metallic mineral deposits such as polymetallic sulphides have instead a complete different genesis. They are concentrations resulting from seafloor hot springs forced outside the seabed crust by submerged volcanic activities in the nearby area (hydrothermal vents) and therefore are said to have *volcanogenic* origin. Other types of mineral deposits such as cobalt-rich ferromanganese crusts originate as a result of metals dissolved in the water column that slowly precipitate (*authigenic* origin) and from hydrothermal mineral-rich fluids that aggregate on the seabed. Deep-sea polymetallic nodules are formed through two different processes: (1) A *hydrogenous* process in which concentrations are formed by slow precipitation of colloidal particles from surface sediments; and (2) A *diagenetic* process (produced by alteration of existing materials) in which concentrations are formed in sediments by supply of ore-forming materials from sediment pore water (ISA 2004). In other words, it is the volcanic activity that provides the source of metals.

From an economic perspective both cobalt-rich ferromanganese crusts and polymetallic nodules contain four elements of economic relevance: manganese, copper, nickel and cobalt. The main use of manganese, copper and nickel is in steel manufacturing. Cobalt finds application in the electrical, communications and engine manufacturing industries and is used as a component of superalloys in civil aviation and in the aerospace industry. Cobalt demand is expected to increase in the coming years, while its supply is uncertain, since it is usually produced as a by-product when mining copper and nickel. Nickel also finds applications in chemical plants, petroleum refineries and motor vehicles.

The ISA estimated the world's potential of underwater minerals for the resources lying in the *Area*. The total value is approximately USD 5.8 billion (ISA 2001a). The research was based on four parameters:

1. A statistical evaluation of known occurrences and reserves;
2. The geologic environment favourable for their formation;
3. Models for sediment type and thickness; and
4. Basement composition.

The potential value of all marine mineral resources lying in the oceans is unknown. The deep ocean crust and the continental crust are made of different types of rock. Geochemical processes of enrichment, sedimentary processes of transport and concentration of minerals differ from land to ocean. However, many processes are common over the entire earth and it is legitimate to speculate that the average mineral content for each square kilometre of seafloor is equal to that of land. Given the ocean's dimension, it is therefore obvious that the riches contained on the seabed and under the ocean floor are far superior to what is found on land.

There are about a dozen types of marine mineral commodities, some of which are presently successfully extracted from the ocean. Those being mined include: oil and gas; sand, gravel and shell; coral; magnesium; salt (sodium chloride); sulfur; diamonds, tin, gold and other heavy minerals. The heavy minerals mined at sea can be classified into three main groups:

1. Sand, gravel and phosphates, including silica sand, aragonite and sulfur;
2. Metallic minerals found in placer deposits[4] such as gold, platinum, tin, titanium and rare earth metals;
3. Metaliferous oxides and sulfides such as manganese, copper, nickel and cobalt found in manganese nodules, cobalt crusts and sulphide-rich mineralization.

The first discovery of a seabed mineralization, which happened by chance in the Red Sea in the 1960s, immediately stimulated the interest of financiers. Another wave of expectations about mining the oceans came in the late 1970s after the discovery of massive sulphide deposits at a Pacific hydrothermal vents site. The initial optimistic resource estimates and the general enthusiasm about the possibility of exploiting marine mineral resources in the near future contributed to generate commercial interest in investors. The international mining industry immediately showed interest in the high grades of the discovered deposits. This paved way to develop the United Nation Conference on the Law of the Sea (UNCLOS) and the International Seabed Authority (ISA). However, expectations were not fulfilled and although throughout the years the technology related to the extraction of minerals from the seabed has improved significantly, marine mining technology is not yet economically viable.

Much of the technological advances that have been developed for the hi-tech offshore oil industry have spilled over into the marine mining sectors. Nevertheless, the commercial exploitation of non-fuel marine deep-sea mineral resources is far from being as well developed. Deep ocean mining is still underdeveloped primarily because of the high cost of production compared to land-based mineral extraction.

Searching and extracting marine mineral resources is more difficult and expensive than onshore. This concerns most of the known marine ores.[5] For many heavy minerals such as ilmenite, rutile and zircon there are land-based reserves that can be presently mined at lower cost. Therefore, marine mining is not competitive when compared to similar land-based projects. Another reason is related to the huge depths involved in mining the ocean. Up until now, the most significant mineralization of metaliferous oxides and sulfides have been found at depths of 5,000 m for manganese nodules and 3,000 m for hydrothermal sulfides. Even though extensive ocean mining is not yet a reality, knowledge and vision about marine minerals is expanding rapidly due to continuous scientific discoveries and technological advances. Consequently, the international mining industry is expected to develop an economically viable technology within the next few decades.

Given the present knowledge of the ocean seabed, the higher potential for marine mining lies in the continental margins because these areas are shallower, contain numerous mineral deposits and are part of the EEZs.[6] Continental margins include an area that is almost 50 percent as large as the land area. A critical issue regarding the exploitation of marine minerals is the potential environmental threat of the mining operations. In order to protect marine habitats and minimize the impact on both the physical and the biological components, the environmental impact must be thoroughly investigated and adequate safeguard devised.

MARINE PLACERS

The term "placer" refers to alluvial deposit of sand and gravel that sometimes can contain valuable minerals such as gold and gemstones (diamonds). Silica from siliceous sand (a.k.a. quartz) is also mined at sea. It is used in metal alloys, plastics, lubricants, in integrated circuits and electrical productions, and by the glass and chemical industry. Silica sand is dredged off Japan, northern New Zealand and in the Baltic.

Sand and gravel

Sand and gravel are mined for two main purposes: construction material and beach nourishment. However, there are other minor uses including landfill, land reclamation (the Netherlands) and carbonate sands for agricultural use as lime (Norway, France and the UK).

The demand for aggregates coming from the construction industry has increased in the last twenty years. The expression 'aggregates' describes a number of materials used in the construction industry, ranging from crushed rock material such as limestone, sandstone, granite and basalt, to naturally occurring sands and gravels. The particles that compose aggregates deposits offshore consist of fragments resulting from rocks located in the adjacent regions and vary widely in composition, grain size and roundness (Fox 1993).

Sand and gravel are the most used aggregates. Sand and gravel are unconsolidated sedimentary material composed of different rock types. The terms "sand" and "gravel" are used to indicate different grain-size classes of rock particles. Basically they just differ in size. There are several scales in use for describing grain size: sand grains, which are generally constituted by quartz, are usually smaller, ranging from 0.0625 mm to 2 mm in diameter. Gravel usually consists of multi-grained rock fragments and includes two classes: "granule" and "pebble", ranging between 2 mm and 64 mm in diameter. Marine deposits of sand and gravel may contain a mixture of grain sizes (Williams 1986). There is very little difference between sand and gravel mined on land and that taken from offshore sites. Since impurities related to marine debris can easily be washed away, the origin of the commodity is negligible for consumers.

Placer deposits of sand and gravel are actually the most widely minerals mined at sea, being dredged all around the world either from the beach or from offshore shallow waters. The worldwide annual value of sand and gravel production is about USD 3 billion, Japan being the main producer. This rising trend in the demand for aggregates has resulted in the growing of the marine dredging industry, which now supplies about 20 percent of the total annual production in Japan, the United Kingdom, Denmark and the Netherlands (Williams 1986). The aggregates industry market is almost always composed of several small firms supplying local markets

(Fox 1993). In the UK marine sands and gravels—with over 20 million tons per year—contribute to about 25 percent of the total aggregate consumption (Fox 1993). In the construction industry aggregates are either used for mixing with cement to make up concrete, or coated with bitumen for road construction.

Beach nourishment is the process of dredging sand from one location and transporting it to another. The purpose is to widen and elevate beaches in order to mitigate erosion and reduce storm and flooding risk. The cost and impact of mining and transporting sand from land-based sources, together with the increasing demand for large beach nourishment projects, have led to a rising interest in offshore sand borrow sources. The offshore sand and gravel industry for beach replenishment is therefore expanding—even though expensive—because it is proving to be the most effective way to handle beach protection. For an economic project to be viable, considerable volumes of high quality matching sand,[7] located close to the beach for nourishment, are necessary.

Usually dredging activities offshore are regulated with the granting of a license for prospecting and extracting specific materials in a defined area.[8] Prospecting for offshore sands and gravels is an activity that typically starts with bathymetric measurements of the seafloor and bottom grab sampling to assess the distribution, thickness and quality of the aggregates. Echo sounder recordings and side scan SONAR techniques are usually employed to evaluate the lateral distribution of the deposits. To thoroughly assess all the parameters required for a sound extraction of the materials, seismic and acoustic prospecting techniques are also used. Even though significant sand and gravel deposits exist in almost all offshore waters, available dredging technology is presently limited to depths of about 30 m.

With regards to production technology, there are two main types of dredging machines: the anchor type; and the trailer type. The former works while anchored, the latter while the ship is underway. Both types of dredges employ huge centrifugal pumps to draw the sand and gravel up through dredger pipes which are up to 1 m in diameter and can weigh as much as 60 tons (Fox 1993).

According to their operational system, dredges can be divided into two main types: mechanically operated; and hydraulically operated. Each type of dredge has its own advantages and disadvantages according to the characteristics of the extraction site. Hydraulic dredges for instance can efficiently dig compacted sediments such as dense sands, gravel, clay and soft rock. Both dredging and sand transport method vary according to the distance between the offshore site and land. Economic considerations of dredging include transportation distances—50 km being the maximum cost-effective hauling distance—and the physical characteristics of the deposit (Williams 1986).

Offshore sand mining for beach nourishment usually employs hydraulic operated dredges, such as cutterheads and hopper dredges. The method used with the cutterhead dredge is to cut the material with a rotating drill, suck the material with high-speed centrifugal pumps and then pump the fluidized mass (or slurry) through a pipeline to the disposal site. The cutter-suction dredge is the most used dredge type. This method is employed for instance when the distance between borrow and nourish sites is less than 5–6 km. If the distance between the two sites is greater, hopper dredges are usually used. Hopper dredges are also deployed for mining offshore deposits of coarse sand and gravel for use as construction aggregate. Since hopper dredges are self-propelled ships, the transport method commonly used is to store the sand in the hopper and then download the material onshore. Discharge may be mechanically operated or accomplished by pumping the material. Usually two twin dredges are employed for this operation—while a ship is dredging the other ship downloads (United States Department of the Interior 2003).

Since sand and gravel mining activities tend to be very focused, site-intensive operations, ocean-dredging processes have a hazardous potential impact that may occur to various marine and coastal ecosystems. Even though the geologic condition of the mining site places constraints on the methodology employed to recover sand, dredging has numerous impacts on geologic processes and conditions, which in turn can affect biological and physical processes. The major biological impact of dredging is the complete removal of the existing benthic community in the area. However, adult organisms from adjacent areas usually recolonize the dredged site after a certain period.

Other potential environmental damages related to dredging include changes in bathymetry, alterations of the bottom substrate, poor water quality, impact on marine mammals, fish, commercial and recreational fisheries and on archaeological resources (United States Department of the Interior 2003). Regarding changes in bathymetry, dredging at a shoal may result in the complete removal of a topographic feature of the seabed. Dredging in flat sand sheets may result in the creation of a new characteristic, such as a pit or a bathymetric depression. At the borrow site, these bathymetric changes can affect marine current patterns, such as strengths and directions of the flow. This, in turn, may modify depositional patterns and have deleterious effects on fish and on marine mammals' populations.

However, the environmental impact on marine organisms of dredging at sea for aggregates is negligible when compared with over-fishing. Sand dredging has also an impact on water quality because of the increased turbidity due to elevated suspended sediment concentrations. This can vary according to the size and distribution of the dredged material, to the type and size of the dredging operations and to water movements at the borrow site. Even though this phenomenon is generally temporary, exposure of marine organisms to turbidity can result in clogging of feeding and respiratory diseases.

During dredging operations, archaeological sites or historic shipwrecks can be damaged or covered by sedimentation. Shipwrecks sites are key habitats for numerous species of fish and are also widely used by sport divers and recreational and commercial fishermen. Consequently, it is crucial that dredging operations must be wisely managed to guarantee that environmental damage to marine ecosystems is minimized. A site screening to determine the most appropriate area for dredging and impact analysis to identify the main environmental issues is therefore always required.[9] Basic information on the biological character of resident benthic communities and the potential rate and success of recolonization should be acquired. The evaluation of the potential dredging consequences on the local wave and current regime should also be assessed. The authority in charge for licensing should limit the maximum depth of excavation in order to prevent alterations in the local bathymetry. A preventive measure to preserve the archaeological heritage is to locate and determine the possible existence of submerged prehistoric or historic sites and shipwrecks in order to protect these resources.

Diamonds

Known diamond placers are located in the continental shelves of South Africa, Namibia and Australia. Regarding the genesis of the African placers, the diamonds have been probably carried to the South Atlantic coast along the course of the Orange River. Long-shore currents, together with wind and wave action, have distributed and sorted the gems in diamond-bearing gravel deposits around the coast (ISA 2001b). The diamond content in offshore deposits varies from 0.07 up to 1.00 carat per million mt.

Diamond deposits where discovered in the late 1950s, however, more thorough explorations were undertaken only in the 1960s. Offshore mining on a permanent basis started in 1989 by Namdeb Diamond Corporation Ltd., a joint corporation owned by the Namibian government and De Beers Centenary AG, while De Beers Marine Company is the contractor for the Namibian waters. Other minor producers are also present in the region. Southern Africa diamondiferous gravels are presently profitably mined and during the whole 1990s the offshore diamond production in the region has steadily grown.

More than 650,000 carats valued at over USD 180 per carat were taken from offshore Namibian mines in 1996, with recovery of 514,000 carats reported for 1999 by De Beers Marine, the principal producer (Jones 1998). Present annual output is about 0.7 million carats and is increasing. Offshore diamond mining off Namibia's (Chameis Bay) and South Africa's Coast (Groen River, Broadacres, Casuarina Prospect) has been considered a 1.5 billion USD per year business (Cruickshank 1998).

Offshore diamond mining involves sophisticated mining operations carried out with different mining systems. Diamonds are either mined by large onshore drills modified to perform marine activity, or by seabed crawlers remotely operated by vessels. These seabed-mining machines have a mass ranging from 50 to 120 tons and dimensions up to more than 20 m. They are track- or skid-mounted and are deployed by vessels. Diamonds are mined with highly selective and aggressive excavation and extraction systems. The underwater device drills and cleans-up irregular bedrocks. Afterwards sediments are removed and subsequently airlifted or pumped, then processed and gems finally sorted.

Diamond mining offshore Namibia is made up of a continuous mining system consisting of 7 m diameter rotating cutter heads attached to a flexible drill string through which the diamond-bearing sediment is airlifted onto the vessel (ISA 2002). The local environment in which the mining activity takes place is very hostile, with strong winds and extreme swells. Therefore, flexible connections with marine risers are required to bring the sediments to the surface.

Diamonds have been successfully mined in heavy gravels on bedrock at depths of 100 m and more. Presently the offshore diamond industry mines the continental shelf using ROVs down to 200 m depths, with leases extending to 1,000 m depth. It costs around USD 80 million to put together a diamond mining system for these depths, compared to around USD 350 million to start a similar operation on land. Offshore production systems seem to be even more efficient than the land counterpart and offshore mining is more profitable than mining for diamonds on land. In addition, the marine system is portable and can be transferred to other sites.

Even though world supply of diamonds totals around 100 million carats per year and therefore offshore production is still small compared with the output from onshore, the onshore production is declining because reserves are exhausted. Moreover, the quality of offshore gemstones is proving superior to those found at onshore sites because of natural sorting processes. The value of offshore diamonds ranges between USD 150–400 per carat. Therefore, although mining operational costs are double than those inshore, diamond producers find offshore production increasingly rewarding.

POLYMETALLIC NODULES

Polymetallic manganese nodules are the result of a slow accumulation process of oxides of manganese and iron and of hydroxides. It has been calculated that they grow by 3 mm per million years (MA), their age having been estimated at 2–8 MA (ISA 2001a). Even though polymetallic nodules contain more than thirty metallic elements, traces of nearly all the elements in the periodic chart can be found. However, they are primarily valued for their manganese, nickel, copper and cobalt content. Nodules contain up to 30 percent of dry weight of manganese (average about 20 percent), up to 17 percent of iron (average about 12.5 percent) and less than 3 percent combined of nickel, copper, lead, zinc and cobalt. Water contained in nodules is about 80 percent of the total mass.

Although nodules' shape vary from nearly flat to spheroidal, ranging in size from 1 to 15 cm in diameter,[10] nodules usually appear as potato-sized spherical concentrations of layered metallic substances lying on the seabed partially buried by sediments. Larger nodules (5–20 cm) are mainly grown by *diagenetic* process and are rich in manganese, nickel and copper but relatively poor in cobalt. Nodules originated by *hydrogenic* process are usually smaller (2–4 cm) and of irregular shape, rich in cobalt but poor in other elements. Although first discovered more than 200 years ago in the Siberian Kara Sea (Artic Ocean), it was since the R/V HMS *Challenger* expedition between 1872 and 1876 that scientists begun to study them extensively. Commercial interest in nodules awakened in the 1960s and 1970s, when huge nodules deposits were discovered and the potential economic value of their strategic minerals content highlighted.

In spite of the well-known origin, mineralogy and composition of the nodules, assessments of their resource potential are not well documented. The currently known distribution of nodules on the ocean bottom is based on information acquired by side scan SONAR, drill cores, freefall grab samples, seafloor photos, video camera records and direct observation from HOVs. Estimates of nodules' total volume vary from the 1965 optimistic figure of 1.5 trillion metric tons (mt) calculated by John Mero (Glasby 2002) to the more realistic figure of about 500 billion mt, of which only 15 to 25 billion mt satisfy current mining criteria[11] (ISA 2001a). An economic criterion to evaluate nodules deposits is to combine their abundance with their metal content. Glasby (2000) calculated that the combined content of nickel, copper and cobalt must be greater than 2.5 percent and the abundance of nodules on the seafloor must be greater than 10 kg per square meter. However, other assessments with this same criterion have been made, resulting in different figures.

All oceans have their own deposits of polymetallic manganese nodules, their distribution density and their metal content and proportion varying greatly according to the host location. Nodule deposits have been found at many depths, however, the highest concentrations lie in the vast plains of the deep seafloor between 4,000 and 6,000 meters depth. The majority of nodules of economic interest occur in the Eastern Pacific Ocean equatorial region, the so-called Clarion-Clipperton Fracture Zone (CCFZ). Estimates for the CCFZ range widely from 1,700 billion to one trillion mt. The most recent resource estimates indicate a nodule resource for this area of 34 billion mt, with a total metal content of about 7.5 billion mt of manganese, 78 million mt of cobalt, 340 million mt of nickel and 265 million mt of copper (Morgan 2000). According to an ISA statistical analysis conducted with its Polydat database, the nodules abundance in the CCFZ has an average of 6.12 kg/m^2. However, nodules concentration of more than 15–20 kg/m^2 is found in many parts of the CCFZ (ISA 2004).

The second largest and richest nodule field in the world is located in the centre of the North Indian Ocean. A conservative resource estimate (Ranadhir et al. 2004) suggests that about a 300,000 km^2 area within this field holds 288 million mt of manganese, 14 million mt of nickel, 13 million mt of copper and 1.7 million mt of cobalt. An estimate for the 150,000 km^2 area allotted by the ISA to India in the Indian Ocean suggests that, being the nodules abundance within a range between 5 to 10 kg/m^2, the resources potential in this area are about 380 million mt of nodules, containing 4.7 million mt of nickel, 4.3 million mt of copper and 0.55 million mt of cobalt (ISA 2004). Another location of interest for manganese nodules is the Peru basin.

Already in the mid-1960s the underwater mining community (five international consortia involving thirty-six companies from seven different countries) began to explore the commercial potential of the nodules. Unfortunately, the excitement did not last long, because of the cost of retrieval and the drop in the demand for steel and nickel. As a matter of fact, it was the great interest that concentrated at that time around nodules that gave birth to Part XI of the UNCLOS and to the ISA.

Harvesting systems to recover polymetallic nodules have been experimented in several areas, such as the region of the Northeast tropical Pacific, the region south of India, the Cook Islands EEZ and the Blake Plateau off the US Atlantic coast. From the 1960s to 1984 more than USD 650 million had been invested to develop new harvesting technologies and to explore for polymetallic nodules (Glasby 2000). However, the financial results of the operations have been nil.

Still today there is no commercial profitability in mining nodules because of their low grade in precious metal content and the costly operations to get them to the surface. Furthermore, there is always a high technical risk of losing the expensive equipment. Nonetheless, deep-sea nodules surely have economic value for future development. It has been calculated that to be commercially viable, a mining venture should have an annual average amount of about 5 million mt of wet nodules for a period of 20 to 30 years (ISA 2001b).

SULPHIDE-RICH MINERALIZATIONS AND COBALT CRUSTS

Before discussing the characteristics of these mineralizations, a brief overview of their genesis must be given. Both sulphide-rich mineralizations and cobalt crusts originate from hydrothermal vents. A hydrothermal vent is a deep seafloor volcanic hot spring spewing a very hot mineral-rich fluid. It resembles a geyser. Hydrothermal vents are usually found at an average depth of about 2,000 m, almost always spreading in proximity of active ocean tectonic plate boundaries, such as along the Mid-Ocean Ridge

Table 5.1 Average Concentration of Metals in Polymetallic Nodules

Element (dry weight %)	*Atlantic Ocean*	*Pacific Ocean*	*Indian Ocean*
Manganese	13.25	20.1	15.25
Iron	16.97	11.4	14.23
Nickel	0.32	0.76	0.43
Copper	0.13	0.54	0.25
Cobalt	0.27	0.27	0.21
Zinc	0.12	0.16	0.15
Lead	0.14	0.08	0.1
Iridium	9.32	6.64	3.48
Uranium	7.4	7.68	6.2
Palladium (parts per million)	5.11	72	8.76
Thorium	55	32.06	40.75
Gold (parts per billion)	14.82	3.27	3.59

Source: UN-ISA (2004) reprinted by permission. All rights reserved.

(divergent plate boundaries) and Volcanic Island Chains (convergent plate boundaries). They form in geologically active areas because of the melting and moving apart of continental plates. The highest level of hydrothermal activity is found on the fastest spreading ridges. The spreading rate of continental plates ranges from 1 to 10 cm per year, depending on the rate of lava supplied from beneath the seafloor.

At divergent plates boundaries, separations create cracks and crevices in the seabed. As seawater seeps into the seabed openings it is heated by the magma beneath, which has a temperature of about 1,200 C° and lies at about 3–3.5 km below the seabed. Once heated and enriched by minerals, the fluid increases its buoyancy and starts to seek its way up to the seabed. When the heated fluid bursts out, its temperature may be as high as 400 C°. However, it does not boil because of the very high pressure found at depths, water pressure at 2,000-3,000 m depth is about 200–300 bars. The heated fluid carries high concentrations of metals, reduced sulphur and other elements that are precipitated when the solution encounters the 2 C° bottom water temperature. The resulting formations are mineral-rich mounds and columns, sometimes rising 9–10 m (and even higher) over the seabed. These columns topping the hydrothermal vents sites are called *chimneys*.

Two different types of hot springs columns are known: (1) The *black smokers*, which spew great amounts of dark particles of iron and sulfides and have a higher temperature; and (2) The *white smokers*, which spew a lighter compound with a lower metal content formed by barium, calcium and silicon. The temperature of the *white smokers*' fluid is cooler than the *black smokers*' one. Vent fluids issuing from hydrothermal vent systems contain abundant methane, but they are also enriched in propane, ethane and many other dissolved hydrocarbons (Foustoukos and Seyfried 2004).

The average size of a hydrothermal vent field is about one hectare and the mineral deposits resulting from the hot springs can reach a size ranging from several thousand to 100 million mt (ISA 2002). Since 1985, when scientists aboard Woods Hole Oceanographic Institute's (WHOI) HOV *Atlantis II* discovered hydrothermal vents at the Trans-Atlantic Geotraverse (TAG) active mound, this is the single largest known active mineral deposit along the Mid-ocean Ridges. The TAG site, located at the Mid-Atlantic Ridge at a water depth of 3,650 m, was drilled during Ocean Drilling Programme (ODP) Leg 158 in 1994 to a total depth of 125 m under the seabed. It has a diameter of about 150 m and rises about 50 m above the seafloor, with an estimated mass of 3 million mt of metals (Kingston Tivey 1998).

Although the occurrence of seafloor hydrothermal vents can be found at different geological environments, higher occurrence and content of valuable metals are mostly found in deposits at Mid-ocean Ridges. The submerged Mid-ocean Ridges strand for some 55,000–60,000 km, encircling the earth at divergent plate boundaries. More than 80 percent of the Mid-ocean Ridges lie in the *Area*. Having investigated systematically only about 5 percent of the ocean floor and less than 10 percent of deep ocean ridges, after 25 years from their discovery it is known that there is about 200 active and inactive hydrothermal vents sites all around the world's oceans bottom. Out of these, only about a hundred sites have been visited thus far and five different bio-geographic provinces have been identified. About 500 previously unknown marine species have been discovered at hydrothermal vents sites. The main characteristics of these organisms are discussed in chapter 7.

Hydrothermal vents sites—active or inactive—contain mineral deposits of two different kinds: polymetallic sulphide-rich mineralization, and cobalt-rich ferromanganese crusts.

Polymetallic sulphide-rich mineralization

Deep seabed sulfide deposits, commonly called polymetallic sulfides or sulphide-rich mineralization, are formed nearby areas of active ocean plate boundaries—both divergent and convergent—including Mid-ocean Ridges, back-arc rifts and seamounts. These are the same environments where hydrothermal vents sites are located. Therefore, sulphide-rich deposits are usually found at or near active or inactive hydrothermal vent sites, as a result of collapsed chimneys accumulating minerals that toppled over. It has been estimated that an average black smoker can produce about 250 tons of massive sulphides per year (ISA 2004).

Polymetallic seafloor sulphides deposits may carry sulfides of almost any metallic element, including copper, lead, iron, zinc, silver and, in lower concentrations, also gold. Their metal content varies, with deposits composed in average by more than 20 percent copper, 20 percent zinc and significant amounts of gold and silver (Jones 1998). Deposits occur in different shapes, such as chimneys, mounds, sediment layers and blocks, and are made of fine layers of different metal sulphides associated with sulphates, silica and clay minerals. Polymetallic massive[12] sulphides deposits have a size between 1 and 100 mt, even though many occurrences amount to less than few thousand tons.

The most significant metallic deposit discovered up to date is Atlantis II Deep, a basin about 2,000 m deep located north in the Red Sea due west of Jiddah, Saudi Arabia. Atlantis II Deep mineralization consists of fine-grained metalliferous mud instead of sulphides. This difference is thought to be a consequence of the high salinity of the Red Sea. Although as early as 1947, Swedish R/V *Albatross* recorded abnormal changes in temperature and salinity at depth in this area, a WHOI's R/V discovered the Red Sea deposits in the mid-1960s. The configuration of this deposit has been subsequently investigated with high-resolution seismic and echo sounding techniques. In 1979 a pilot mining test conducted by the German Company Preussag for a Saudi-Sudan joint commission (Atlantis II Deep is located in their EEZs) has shown that the deposit can be successfully mined with water jetting and hydraulic pumping techniques. The 40-km^2 Atlantis II Deep deposit has a potential of metallic sediments of about 100 million mt, containing a significant grade of zinc (2 percent), copper (0.5 percent) and silver (38 gr per mt). Atlantis II Deep was estimated to have a value of about USD 5 billion in 1985 prices (Glasby 2000). The significance of the potential resource of the whole area is even bigger, since Atlantis II Deep is a just one out of 21 deeps located in the Red Sea. However, the study concluded that the mining venture would not yield the required returns (ISA 2002).

Another significant discovery was made in 1991 at hydrothermal vents site off the northeast coast of Papua New Guinea's territorial waters in Easter Manus Basin (Pacmanus). This remarkable sulphide-rich mineralization deposit, which lies at a water depth of about 1,600 m, gave way to the first exploration licence issued in 1997 by New Guinea's authorities to an Australian mining company, the Nautilus Minerals Corporation.[13] Neptune Resources, another Australian mining company, has recently obtained a licence to explore a hydrothermal field off the North coast of New Zealand. Both sites have high gold and base metal grade. Large sulphide deposits had also been discovered in sediment-filled basins, such as the Guaymas Basin in the Gulf of California and the Middle Valley deposit on the Juan de Fuca Ridge offshore Canada, as well as in back-arc spreading centres, such as the Central Manus Basin and the Mariana Trough.

In some ways, the polymetallic hydrothermal sulphides look a much easier exploitable venture than mining manganese nodules, since the grade of the ore is higher and the minerals rest in shallower water—on average about 2,000 m compared to 5,000 m of the nodules. However, only a dozen deposits (out of which two are located in the *Area*) have the sufficient

Table 5.2 Average Composition of Polymetallic Massive Sulphides Deposits

Element (dry weight percentage)	*Mid-Ocean Ridge*	*Volcanic Island Chains*
Lead	0.1	6.1
Iron	26.4	9.6
Zinc	8.5	18.35
Copper	4.8	3.65
Barium	1.8	9.9
Arsenic (ppm*)	235	9,172
Antimony (ppm)	46	3,408
Silver (ppm)	113	1,260
Gold (ppm)	1.2	3.8
Number of samples analyzed	1,259	613

*ppm = part per million (equals 1 gr per ton).
Source: UN-ISA (2004), reprinted by permission. All rights reserved.

size and grade to be considered economically exploitable for future mining, even though their deposits' thickness is still unknown.

Cobalt-rich ferromanganese crust

Cobalt-rich ferromanganese crust is a mineralization that is also commonly called cobalt crust, ferromanganese crust, manganese crust, or high-cobalt-manganese crust. This kind of deposits were first discovered by chance in 1978 by a French/Mexican/American expedition at East Pacific Rise, a submerged volcanic mountain lying south of the Gulf of California. Ferromanganese crusts are the result of precipitated metals dissolved in the water and discharge fluid from deep-sea hydrothermal vents hot springs. Like the nodules, they are relatively rich in cobalt, nickel, manganese, titanium, platinum group elements[14] and other metals. Cobalt crusts occur throughout all oceans and are usually found on hard-substrate seabed elevations, such as seamounts, ridges and plateaus. They precipitate from cold ambient seawater onto rock substrates but instead of accumulating in nodules lying in the seabed, this mineralization occur in encrustations adhered to volcanic rocks. Encrustations can be up to 50 cm thick, more commonly 3–5 cm. They occur in water depths between 400 and 4,000 m and are primarily evaluated on the basis of their cobalt content.

To evaluate cobalt-rich ferromanganese crusts, one of the conditions is that the crust should be at least 40 mm thick and contain on average more than 1 percent of cobalt content (Glasby 2000). The economic value of cobalt crusts is therefore a function of the cobalt price. Cobalt is used to make corrosion-resistant, light and strong high temperature metal alloys

Table 5.3 Range of Average Concentration of Metals in Cobalt-Rich Ferromanganese Crusts from the Pacific, Atlantic and Indian Oceans

Element (dry weight %)	*Range of concentration*
Iron	15.1–22.9
Manganese	13.5–26.3
Nickel (ppm)	3,255–5,716
Copper	0,713–1,075
Cobalt	3,006–7,888
Zinc	512–864
Barium	1,494–4,085
Molybdenum	334–569
Strontium	1,066–1,848
Cerium	696–1,684

Source: UN-ISA (2004), reprinted by permission. All rights reserved.

and paints. Having as well superior magnetic properties, cobalt is also used in metallic forms as an essential component in permanent magnets. The formation process of these crusts takes a million years, since they generally grow 1–6 mm per MA. Even though ferromanganese crusts occur in every ocean at different latitudes, the most detailed studies have analyzed the equatorial Pacific site. Pacific deposits have the highest mean cobalt content and thus they look like the most promising. According to one

Table 5.4 Value of Metals per Metric Ron of Cobalt-rich Ferromanganese Crust from the Central-Equatorial Pacific

Element	*Mean Price 1999 ($/kg)*	*Mean Content in Crusts (ppm)*	*Value per Metric Ton of Ore ($)*
Cobalt	39.6	6899	273.2
Titanium	7.7	12035	92.67
Cerium	28	1605	44.94
Zirconium	44.62	618	27.58
Nickel	6.6	4125	27.23
Platinum	13,024	0.489	6.37
Molybdenum	8.8	445	3.92
Tellurium	44	60	2.64
Copper	1.65	896	1.48
TOTAL			480.03

Source: ISA (2002), reprinted by permission. All rights reserved.

Table 5.5 Average Concentration of Valuable Metals in Polymetallic Nodules and Crusts from Different Environments (percentage of dry weight)

Element	*Seamounts*	*Plateaux*	*Active Ridge*	*Other Ridge*	*Continental Margin*	*Marginal Seamount*	*Abyssal Plains*
Manganese	14.62	17.17	15.15	19.74	38.69	15.65	16.78
Iron	15.81	11.81	19.15	20.08	1.34	19.32	17.27
Nickel	0.351	0.641	0.306	0.336	0.121	0.296	0.54
Cobalt	1.15	0.347	0.4	0.57	0.011	0.419	0.256
Copper	0.058	0.087	0.081	0.052	0.082	0.078	0.37
Mn/Fe	0.92	1.53	0.8	0.98	28.8	0.81	0.97
Depth (m)	1,872	945	2,870	1,678	3,547	1,694	4,460

Source: ISA (2001a), reprinted by permission. All rights reserved.

estimate (UN-ISA 2004), cobalt-rich ferromanganese crusts cover 1.7 percent of the ocean floor, total cobalt content being about 1 billion mt. The Central Pacific Ocean alone is supposed to contain about 500 million mt of cobalt.

MINING TECHNOLOGIES AND ENVIRONMENTAL IMPACT

Marine metal deposits knowledge and mining techniques evolve according to advances in research and technology. Over the past forty years researchers have drilled as many as 1,700 holes per year in the ocean floor and studied more than 35,000 samples. Since 1984 this work has been carried out under the 22-country ODP, which disposed of a R/V equipped with borehole tools and other devices applicable to seabed borehole mining. ODP's annual budget was USD 80 millions, 60 percent of which provided by the US National Science Foundation (Normile and Kerr 2003). The ODP has been recently replaced by the Integrated Ocean Drilling Program (IODP), being carried out by Japan, the United States and the EU.[15] IODP is a multi-platform marine research program that involves two drilling vessels and some other mission-specific platforms. This new initiative was launched in 2003 and is a 10-year 1.5 billion USD project. Its centrepiece is the drill ship *Chikyu* provided by JAMSTEC (Japan Marine Science and Technology Centre), which houses the world's largest deep-sea drilling, a device capable of boring nearly 7 km into the ocean floor.

Regarding mining terminology, even though the UNCLOS does not include this difference, it is common and international usage to distinguish between reserves and resources. This distinction is similar to the one used to estimate the amount of hydrocarbons, being mainly based on the economical recoverability of the material. Reserves comprise known abundance and volume of marketable and economically recoverable materials. Resources comprise estimates of the potential occurrence and abundance without tacking into account the commercial feasibility of their exploitation.

Another important distinction is the one between "prospecting" and "exploration". Prospecting, which is the search for mineral deposits including estimates of their composition and value, is usually based on the ability to predict the environment of potential metal formations—i.e. the study and activity of searching for favourable target areas. Exploration is the searching activity related to collecting and processing samples, as well as analysis and tests of equipment.

Regarding mining activities in the *Area*, the ISA regulation for polymetallic nodules—the so-called 'Mining Code'—provides different rules for prospecting and exploration. A prospector seeking an exploration contract from the Authority must submit an application that "...shall cover a total area, which need not be a single continuous area, sufficiently large and of sufficient estimated commercial value to allow two mining operations. The application shall indicate the coordinates dividing the area into two parts of equal estimated commercial value..." (UNCLOS, Annex III, Art. 8).[16] Within a prescribed period of time, the Authority designates one of the two areas identified for mining operations as reserved for exploration either by its operating arm (the Enterprise) or by the Authority in association with developing states. The second of the two areas identified for mining operations is then allotted to the applicant. In the ISA regulating framework, prospecting does not confer exclusive rights, while exploration confers exclusive rights in a delimited geographical area.

The exploration phase aims to find out the exact location of a mineral deposit and to determine its physical and chemical properties. The exploration process usually consists of successive evaluations of increasing sophistication and details of an area, eliminating portions without potential mineral occurrence from the initial area of interest. Usually, the search starts within a wide seafloor region in which that specific kind of mineral deposit is known to take place. Components of the mineral ore can be present even at very long distances from the source in the form of particles dissolved in the water column. Therefore, researchers follow the concentration gradients both in the water column and in the seabed sediments by means of water sampling techniques. Researchers employ bathymetric and magnetic methods and acoustic techniques such as side scan SONAR in order to progressively restrict the area of interest. Once the field of exploration has been narrowed enough, imaging methods from UUVs and direct observation from HOVs are finally employed to study the ore in detail. Having identified the targeted area, the mining company wants to know the mineral occurrences at the deposit to determine the value of the findings.

The two most used techniques in marine mineral exploration are bottom sampling and acoustic sounding. Acoustic sounding techniques have been extensively explained in the second chapter. Bottom sampling devices include free-fall grabs, a variety of corers (such as spade, piston and gravity corer) and dredges (such as barrel, box and chain-bag dredge) as well as TV-guided grab systems. A free-fall grab consists of a float, a grab and one or more ballasts. Once launched into the water it sinks because of its own weight (around 20 kg). When it reaches the seafloor it penetrates the bottom sediment and the lever arm closes, trapping the sediment material. Then ballasts are automatically released causing the float to raise the grab. Once back at the surface the grab is located and retrieved. A spade corer is instead lowered into the water by a winch. It contains a sampling box coupled with a spade trigger that penetrates into the seafloor sediment to store the material. Afterwards, the spade corer is lifted back to the surface. A piston corer is used when a long sediment sample has to be collected. It consists of a long heavy core tube lowered into the water by a winch that penetrates the soil by means of its own weight. Dredges bottom samplers are a wide class of device suited for both soft bottoms and hard samples. They usually consist of a long hollow metal pipe or alternatively a box lowered to the seabed to collect material. Regarding bottom sampling a mineral deposit, it is generally agreed that samples dredged by vessels or collected on the seabed by underwater vehicles do not produce the necessary information to completely assess the continuity and distribution of a deposit, but only its surface appearance.

The sampling methods mentioned above can provide information on the two horizontal dimensions of the mineral deposit—i.e. that part of the deposit that is exposed on the seabed. Since metals are usually unevenly distributed in the deposit, drilling for samples' recovery and analysis is the main operation to assess metal content and evaluate reserves. Regarding the third dimension—the depth—only indirect information can be acquired, as this part of the deposit lies beneath the seabed. Therefore, drilling is necessary in order to assess precisely grade (metal content) and tonnage (amount of sediment containing concentration of metals) of the mineral deposit.[17] Even though necessary, drilling and recovering of corers at sea is a very challenging, costly and difficult activity.

Recently, in addition to classic drilling vessels, a portable seafloor drilling and coring system called PROD (Portable Remotely Operated Drill) has been developed by the US-Australian company Benthic GeoTech Pty Ltd together with Williamson and Associates Inc. in collaboration with the University of Sydney. Its performance specifications, at an operating depth of down to 2,000 m, include a maximum penetration of 100 m, a core diameter of 40 mm and core length of 2.2 m. Another drill called BMS (Benthic Multicoring System) has been developed by an Australian/Japanese joint venture and built by Williamson and Associates Inc. for the Metal Mining Agency of Japan. This device uses conventional diamond rotary-rock and soil sampling tools and has a maximum operating depth of down to 6,000 m. Individual corers have a diameter of 44 mm and a maximum coring depth of 20 m (ISA 2002).

Retrieval technology includes devices to collect materials from the seabed and lift it to the surface. These operations have to take place in a very difficult environment, since mineral deposits can be located in an uneven seafloor at great depths (in the order of thousands of metres) and thousands of kilometres away from the coast. In addition, high waves and frequent storms make the stabilization of the mining platform a big issue. However, the main challenge in marine mining is still the retrieval technology, which is high in cost and time and therefore inefficient. Most of the projects are technically feasible but not economically viable. Uncertainty and high risk, coupled with high costs, still influence and prevent investment decisions in seabed mining. A number of devices have been tested in the real deep-sea, some only in the laboratory, while others have not yet been tested at all. During the 1970s, an international team (Ocean Management Inc.) conducted a pilot mining test program in the Eastern Equatorial Pacific and successfully mined nodules proving the technical feasibility of deep-sea mining together with the uneconomical viability of the venture.

There are three main basic technologies for deep-sea mining (Krutein 1980): hydraulic; pneumatic; and mechanical. They all have three main common features: (1) the mining system aims at separating as much unwanted material as possible just on the seafloor, usually through a crusher system such as hammer crushers and ball mills; (2) they are composed of a floating platform or a mother ship to store the material, by submerged equipment to collect the underwater resources and by a double connection system (one for control and one to raise the material); (3) the entire system is portable. All types of underwater mining systems share a common benefit over similar land-based operations, i.e. the start-up costs are just one-time costs. When companies finish mining an area, they simply take their equipment and move to another site and there is no infrastructure left behind to decommission.

The submerged equipment and the riser are crucial components. Concerning nodules, several mining devices have been developed in the past and some are still in the developing process nowadays. The submerged equipment to pick up polymetallic nodules can be: a passive towed mining device dragged by a ship such as dredges or sledges; or an active self-propelled mechanical crawler that moves on the seabed. Compared to passive devices, active mechanical devices are more complex, heavier and require their own power, and are therefore less reliable for commercial operations. However, active devices imply increased control and thus a more efficient utilization of the mine site, coupled with a better manoeuvrability for obstacle avoidance and less risk of clogging. Therefore, good horizontal and vertical manoeuvrability, higher recovery rate of resources and less seafloor disturbance suggest that self-propelled vehicles are highly recommended.

The active mining vehicle includes a mechanical collector that usually has a comb-like structure similar to a potato-harvesting machine. Hydraulic suck-up devices and hybrid hydraulic-mechanical collectors have been tested as well. Slurry pumping—the main type of hydraulic lift—has the advantage of being a simple device without sophisticated components that possesses high lift efficiency. On the other hand, hydraulic collectors have been proved to have the disadvantage of a very high-energy consumption. Mechanical collector devices are also under development.

The Institute for Construction of the University of Siegen, Germany (IKS), is developing an advanced active mechanical nodules collector that consists of a pick-up device able to clean the nodules, a steel belt conveyor and a crusher (ISA 2001b). Telemetry and remote-control system to manoeuvre the collector are integral parts of any mining device, since the collection process must be monitored and managed from the mother ship. A remote-control system consists of a controller, an underwater measurement and control-unit, and a power and communication cable between the submerged equipment and the monitoring station at surface.

The riser system is the equipment used to lift the material to the surface. It can be an umbilical made up of a long lift pipe (either hydraulic or pneumatic), or a mechanical system like the classic dredging. Recently developed lift pipes are composed of a steel part attached to the mother station at the surface (a drilling ship or a barge) and a flexible hose attached to the collector (ISA 2001b). These pipes are similar to marine risers used for deep-water offshore oil and gas extraction. The flexibility of the hose allows a certain freedom of movement to the mining machine, thus enhancing the effectiveness of the whole system.

The continuous line bucket system has also been proposed and successfully tested for mining nodules and cobalt-crusts. This device consists of dragline buckets connected together on a cable loop long enough to extend from the ship to the ocean floor. The cable continuously rotates so that each bucket travels in turn to the seabed, collects nodules together with sediments and then surfaces to deposit its content on the ship. The bucket mining system is simple, effective and involves comparatively low operating costs. However, it has been abandoned since it has a violent environmental impact especially on the soil and water column.

Once metals are collected they must be processed. Metal processing technology has two main systems: pyrometallurgy, or smelting; and hydrometallurgy, or leaching. The pyrometallurgical process consists of smelting the mineral material to a crude alloy and then extracting the metals by use of chemical additives. The hydrometallurgical process consists of separating and discarding sediments to extract the interested metals by breaking down the matrix mineral by chemical reduction. Usually ammoniacal sulphur dioxide, reduction roast ammonia, or sulphuric acid-starch leach processes are employed.

Hydrometallurgical treatments offer a higher degree of selectivity together with a greater flexibility during the operations. Moreover, leaching permits to recover a greater number of substances. Processing costs are largely proportional to the volume of mineral processed. The plant may be based onshore or on a floating platform located at the surface above the mine site. When the plant is based onshore, sometimes a pre-concentration process of the metals is performed aboard the vessel in order to avoid both high transportation costs and storage and disposal problems onshore.

ECONOMIC FEASIBILITY OF MINING THE DEEP OCEAN

To place a monetary value on deep-sea mining projects is a very difficult task. Generally speaking, mining is a high-risk venture in itself because of a number of unpredictable factors related to the finding, acquisition, extraction, treatment and sale of the product. Risks may be of a geological, technological, economic, political and social nature. In addition the whole mining activity—exploration, prospecting, discovery, evaluation, extracting and processing—depends upon sale of the product within a long preproduction and production frame time. The economic feasibility study of a mining venture is an assessment of the relative viability of the project involving a review of: the potential revenues and the costs.

Revenue is a function of both the recovered resources and the market value of the extracted metals. Resource estimates have to be based on predictions of the physical and chemical characteristics of the mineral deposit. This can be done by collection of samples and analysis, and through modelling of the predicted tonnage (quantity) and grade (quality) of the deposit. To forecast tonnage and grade, a number of physical characteristics of the deposit an assessment is necessary. This includes size, shape and continuity of the mineralized zone as well as frequency distribution and the spatial variability of the metal grade. Stochastic variability of the mentioned parameters is taken into account in order to complete the information requirements. Apart from the great probability of sampling errors, even core drilling can be affected by errors. Although core drilling is of fundamental importance in determining the physical characteristics of the deposit, it is not certain that the same ore-grade of adjacent cores is to be found in between. In addition, resource estimates must include some knowledge of the proposed mining technology, since different mining methods probably affect size, shape and grade of the potential reserves. The degree of accuracy of resource estimates provides the basis for financing the mining venture and ensures the profitability of the project. Usually resource estimates are far from being precise even for land-based mining ventures. Therefore, even more imprecise resources estimates are to be expected for deep-ocean projects.

Both supply and demand of minerals are cyclical and variable. Consequently, prices of metal and mineral commodities fluctuate widely over time as a result of major factors including economic, technical, social and political ones. Technological advances, new discoveries, substitution by other commodities, fashion, wars and embargoes, development of new industrial uses, are all factors that influence metal prices. Usually mineral material can be sold at different stages of development, such as concentrate, intermediate metal, refined metal, alloyed metal, semi-finished product, finished products, or as scrap. The prices of metals are generally set internationally. Some metals have an established terminal market. The most important international terminal market for base metals is the London Metal Exchange (LME). The major terminal market for copper, silver and gold is the New York Commodity Exchange (COMEX). Quotations from these markets are the basis for metal transaction prices all over the world. Due to the high degree of internationalism in the minerals commodity market, a shortage or surplus in one part of the world can influence the market worldwide. Market prices may be depressed by both the discovery and development of large or high-grade deposits and by substitution of alternate materials. Since the magnitude of ocean mining could be enormous, the amount of metals produced from a single major operation could alter world prices. World metal prices have been depressed since the beginning of the 1980s and consequently most of the land-based mines do not work at their full capacity.

Costs include capital investment and operational costs. Mining ventures are capital-intensive investment faced with costly environmental problems. The two most important factors affecting operational costs are the size of the deposit and the extraction and processing technology. Therefore, costs are sensitive to the accuracy of the resource estimate. In estimating the economic viability of marine mining projects, the main challenge seems to be predicting costs. Although it is possible to estimate the value of a unit of *in-situ* ore of a marine mineral, it is very difficult to assess deep ocean mining production costs and, consequently, it is impossible to produce reliable economic feasibility studies. All in all, the commercial feasibility of mining the seabed for minerals is difficult to assess. Production costs of marine mining projects are plagued with uncertainties: costs vary from deposit to deposit; from one mineral to another; and from one technological system to another. Obviously, they escalate with the increasing depth. In addition, production processes and underwater technologies bear an additional high degree of risk because of the harsh and unpredictable marine environment. Under these circumstances, economic feasibility studies cannot be carried out in a meaningful way and therefore, even if technically feasible, the profitability of marine mining cannot be determined. In economic terms mining the deep oceans for minerals is still a futuristic venture. However, a general overview of present and future marine mining activity can be given.

Some minerals are successfully mined at sea, namely minerals found at placer deposits. The main reason is that placer deposits are located either in marine beaches or near-shore and thus they are more accessible than deep-sea minerals. Tin, which is found in cassiterite placer deposits, is mined in South-East Asia offshore Thailand (Thai Muang, Tongkah Harbor and Takua Pa), Myanmar (Heinze Basin) and Indonesia (Copat Kelabat Bay, Laut Tempilang and Belitung). Tin is used for the fabrication of containers, solders, engine bearings, and air-cleaner and oil-filter cartridges. Offshore production of the mineral almost ceased in Malaysia and Thailand after the tin price collapsed in 1985. However, Indonesia still gets half of its tin production from the sea. Tin-rich gravel is dredged from water depths down to 50 m. Offshore tin production accounts for about 10 percent of the world's production, with an annual value of about USD 100 million (UN-ISA 2004). However, the tin market is presently oversupplied and, therefore, the potential for offshore mining in the near future is minimal.

Gold is mined in Alaska (Nome and Bluff Solomon) and New Zealand (Gillespie's Beach) from time to time according to its market price. In Alaska between 1987 and 1990 total production of gold was 3,672.6 kg, extracted from 4,460,000 m^3 of sediment with a gold content of 0.82

gm per ton. According to UN-ISA (2004) taking the actual gold price, a deposit to be commercially exploitable should have a gold content of at least 2 gm per ton. In 1997 the Australian firm Nautilus Mineral Corporation won a licence to mine gold from an inactive hot vents site in the seabed off the coast of Papua New Guinea, at 1,500 m depth. Other heavy metals are mined in India, Alaska and Madagascar.

Deep-sea deposits that over time have generated interest in the mining industry are manganese nodules and crusts, polymetallic sulfides, phosphates and methane hydrates. Since the mid-1960s there has been commercial interest for mining polymetallic nodules, especially after the 1978 invasion of the Shaba province of Zaire by insurgents from Angola and Zambia. As the ex-Shaba (now Democratic Republic of Congo) is the world's largest producer of cobalt, there had been at that time a rally of the cobalt price, thus fueling interest in alternative sources.

Concerning the *Area*, significant resources of a wide range of minerals do exist and are becoming more attractive because of new cheaper and more efficient marine technologies in exploration and production operations. However, polymetallic nodules still remain the most promising type of mineral deposits in the *Area*. The ISA has issued in 2001–2002 contracts for exploration in the *Area* to seven registered pioneer investors. Six pioneer investors out of seven applied for exploration rights (23 blocks) within the nodule-rich CCFZ in the Pacific Ocean, an area approximately 3,200,000 km^2 wide —i.e. not much smaller than Europe—with depth ranging from 3,500 m to 5,500 m. Regarding mining ferromanganese crusts, this is technologically much more difficult than mining nodules, as the latter lay on a soft-sediment substrate, while the former are attached to substrate rock (ISA 2002). All potential hydrothermal mineralization currently known—except for the sites at the East Pacific Rise at 13°N and TAG hydrothermal field—are located in the EEZ of coastal states (Saudi Arabia, Sudan, Canada, Ecuador, Papua New Guinea, Tonga, Japan and Fiji). With regards to sulphides mineralization occurring at hydrothermal vents fields, the economic feasibility of mining the ore is a function of a number of characteristics of the deposit. These include the morphology of the structure where the mineralization occurs, the average thickness and the internal structure of the deposit, and its metal content. A marine mineral sulphides deposit is considered feasible for mining if it has characteristics such as: (A) inactivity and sediment-free, so as to provoke the minimum environmental impact; (B) sufficient size and tonnage with high base metals and gold grades, since mining higher grades means processing less rock; (C) proximity to land; and (D) a water depth not exceeding 2,000 m (ISA 2004). Since samples dredged, grabbed, or even collected from underwater vehicles produce only information related to the surface of the deposit, recovery and analysis of cores is crucial to determine grade and tonnage of a deposit. As already mentioned, to recover cores from the depth is still a costly technological challenge. Design and construction of mining systems for sulphides deposits are a major task for a number of research institutes and offshore industries. Garnet (ISA 2001b) has proposed to convert the continuous mining system used by De Beers Marine to mine diamonds offshore Namibia and employ it in mining minerals. Grinding with a large rotating cutter head would allow to crush sulphides and altered rocks, and to convert the material into slurry to be airlifted to the surface.

Although still expensive, marine mining technology is developing constantly. In the last thirty years more than 100 deep-sea mining designs (mainly related to polymetallic nodules) have been patented all around the world (ISA 2001b). Newly discovered mineral deposits and recent offshore diamond mining successes offer encouraging results for new research. Results of several research projects developed in the area along the South Brazilian Coastal Zone mention occurrences of deposits of heavy minerals, bioclastic carbonate, quartzose sand, phosphatic concretions and peat (Soundings 2003).

Regarding conservation of marine life, it is clear that mining has destructive effects on the seafloor environment at the mine site and considerable effects on the adjacent water column. The magnitude of the impact depends upon the technology employed. Discharge of bottom sediments in the water column causes the formation of a large plume of suspended material that—according to tides, currents and winds—can extend miles from the mining area. Suspended particles can affect light conditions and chemistry of the water column in areas even far from the mining site. According to Thiel and Tiefsee-Umweltschutz (2001), when polymetallic nodules are sucked up from the seabed and transported to the surface, sediment clouds as large as 7,400 tons are generated daily through the mining process. Moreover, bringing up to the surface a large mass of water from the deep seafloor would change the water composition of the upper levels of the water column, interfering in the equilibrium between surface-deposit feeders and suspension feeders. Eventually, there will also be re-sedimentation effects on other areas of the seabed.

The environmental aspects of marine extraction activity must be carefully taken into account in prospecting, exploration and exploitation of marine minerals, assessing disturbance at extraction sites and the effects to adjacent areas. In order to assess the predicted environmental impact, the major physical, chemical and biological factors that control structure and function of the marine ecosystem in the mining area must be identified before mining operations begin. In accordance with the UNCLOS and the Agreement relating to Implementation of Part XII, the ISA has established in 2001 some environmental criteria and guidelines for prospectors in the *Area*: "Recommendations for the Guidance of Contractor for the Assessment of Possible Environmental Impacts from Exploration for Polymetallic Nodules in the Area." However, it is the contractor who is required to establish environmental baselines against which to assess the likely effects of the program of work. Coastal states also have their own rules and regulations for mining their own EEZ. Regarding regulation, efforts should be made to minimize the environmental disturbance without at the same time paralyzing or inhibiting the development of deep-sea mineral exploitation. Technology should meet the requirements for combining mining and preventing harmful effects on the marine environment or—even better—exploiting at the same time the associated biological resources. In particular, the numerous marine bottom-living organisms living in the mineral deposits' area should be protected or used.

Marine mining activity can conflict not only with the ecosystem equilibrium, but also with the exploitation—actual and potential—of other marine resources useful as biogenetic materials. There are numerous biological resources living in the areas where mineral deposits are found. These organisms, such as for instance *extremophiles* (bacteria living in the deep seafloor) and *thermophiles* (the hydrothermal organisms) have economic value because they hold potential for biological products (enzymes) of great utility.[18] To minimize the disturbance effect on the ecosystem, two main suggestions to operators are commonly agreed upon. One regards discharge of materials, which should be done at a deep level, where currents are very slow. The other one is to leave some preservation zones throughout the mining area in order to prevent species extinction.

Studies have been conducted on reproductive cycles and re-colonization potential of hydrothermal vent communities. Judging from the rapid colonization of new vents following seafloor eruptions, it appears that vent communities are very resilient and have the ability to re-establish within a couple of years even at a severely disturbed site, as long as there are hydrothermal emissions to support microbial chemosynthesis (ISA 2004). Experiments to assess biological recovery and re-colonization after disturbance have been conducted quite extensively. It seems that the time populations need to return to their original level is about seven years. However, the greater intensity of the disturbance from large-scale mining is expected to lengthen the biota recovery times (ISA 1999). Almost all scientists involved in predictions about the adverse mining impact on deep ecosystems express the need to obtain a better understanding to develop effective guidelines on monitoring and managing the deep ocean environment.

6 Living Resources

BIOTA AND THEIR HABITAT

The oceans harbour an ecosystem whose biomass[1] is far larger than that of terrestrial biota. Approximately 300,000 organisms inhabit the oceans (Malakoff 1997); among them more than 20,000 are fish (O'Dor 2003). However, just a fraction of the known marine species has been scientifically named and taxonomically identified[2] because of classification problems with the identity of the specimens. In fact, only some species' names are well established scientifically and marine biota classification problems are well acknowledged among experts.[3] The National Marine Fisheries Service (NMFS)—the Agency in charge of managing the U.S. living marine resources—owns a database of biota in which information about organisms is rated with confidence limits on scientific reliability (Mickevich and Collette 2000). Estimates of the total number of marine living species vary between 500,000 and 10 million (Malakoff 2000). Microbes (such as bacteria and single celled protists/eukaryotes), whose species diversity is unknown, are excluded. New species, including corals, fish and plants are discovered during each new expedition in the underwater world. Small drifting planktonic organisms photosynthesize all the primary food and make up 99 percent of the 145 trillion tons of estimated marine biomass (O'Dor 2003). Although these microscopic factories account for almost all biomass and are the essential catalysts for all biogeochemical cycles in the ocean, the focus of this research is only on commercially valuable species—i.e. finfish and shellfish.

Approximate calculations of commercially valuable species volumes and distributions have been done for certain coastal regions of the world. However, no one really knows how much the entire biomass of economically valuable marine biota is and how it is distributed spatially. This is especially true for the open seas. Continental shelves, which represent only 10 percent of the ocean area, contain most of the known marine biodiversity. As they provide breeding and nursing places for marine life, coastal areas host approximately one-third of the biota. The rest is high-sea fish. FAO—the UN organization in charge of living marine resources—lists about 400 species of marine creatures that can be considered as oceanic or high seas (*pelagic*). For megafaunal species—i.e. organisms that can be seen with the naked eye—there is a clear separation between *bathyal* (continental slope) and *abyssal* species.

The continental slope is a highly heterogeneous environment, ranging from mountainous rocky slopes crossed by deep canyons to lightly inclined, sediment areas made by gravels or mud. Therefore, large-scale habitat variability is found in these areas. On the other hand, species that live in the abyss are mainly small organisms such as *protists* (minute single-celled eukaryotes), crustaceans and worms. Apart from hydrothermal vents' fauna, larger animals like fish are quite rare in the abyss. A number of key biophysical factors such as temperature, pressure, and food availability change with increasing depth, affecting the local environment and consequently the species of hosted biota. Hence, small-scale habitat variability is found in the depths of the oceans.

Life in the oceans is dominated by physical factors such as winds, currents, tides, eddies, waves, salinity, temperature, and photosynthesis, which influence dynamics and variability of chemical, biological, and sedimentological processes. Ocean waters are stratified according to temperature, salinity, and density. Consequently, classical physical oceanographic research has always conducted traditional measurements of conductivity, temperature, and depth (CTD). Other parameters commonly analyzed today are pH (acidity-alkalinity), Eh (standard oxidation-reduction potential), DO (Dissolved Oxygen), nutrients, organic carbon, suspended particulate matter, and heavy trace metals.

Water stores heat more effectively than air and, as a result, temperatures fluctuate much less in the ocean than in land. For this reason, marine organisms generally have a narrower range of temperature tolerance than terrestrial ones. Ocean salinity is generally around 35 parts per thousand, being lower in estuaries and higher in the deep seas. Measuring currents is a fundamental practice of physical oceanographers. By determining how ocean waters move, scientists can determine how organisms, nutrients, and other biological and chemical constituents are transported throughout the ocean. Currents keep the ocean water in constant movement. The complicated pattern of currents is affected by the rotation of the Earth and its major landmasses, the energy from the sun, the bottom topography of the oceans and atmospheric changes such as winds, water temperature, and salt content. Ocean circulation pattern contributes to the heat transport from the tropics to the poles, helping to equalize Earth's surface temperature. The largest ocean circulation pattern has been compared to a global conveyor belt circuit. The cool, dense water sinks near the poles and circulates throughout the deep oceans to surface in the Indian, Atlantic and Pacific Oceans. Then it travels along the surface back again toward the poles, where it sinks once more. Cold deep waters hold a greater amount of oxygen and essential nutrients than warmer surface waters and this pattern of circulation helps in distributing this wealth to marine biota. From an economic point of view, one of the most important features of oceanic circulation is that it is responsible for areas of high fish productivity. The moving mass of nutrients-rich deep water wells up at certain points along a number of continental coasts. These 'upwelling' points are the regions of highest biological productions. In addition to their important role in the global transfer of organisms, nutrients, and sediments, ocean surface currents have a great effect on weather pattern around the globe. On the other hand, deep-ocean circulation is mostly driven by thermohaline effect, which produces differences in the density of water masses.

Two strong constraints on deep waters circulation are: (1) the vertical component of the Earth's rotation, and (2) the topography of the ocean basin. These two constraints are merged as PV (Potential Vorticity) and measured by the ratio H/f (H= ocean depth; f= Coriolis parameter) (Van Dover 2002).

Physical factors and hydrodynamic processes, which take place at various scales and have high spatial-temporal variability, in turn determine the nature and character of biological communities. Marine organisms also depend on the relationship between biotic and non-biotic factors and on the degree of productivity of the marine environment. The productivity at sea is based on concentration of *phytoplankton* (i.e. microscopic plankton functioning just as plants) in open seas, and minerals and organic matter carried away by water running off the land in areas closer to the coast. These factors in turn determine concentration and dispersion of species in different areas. However, there are no clear-cut boundaries.

Productivity varies greatly in different marine areas but by far the most productive ecosystems are coral reefs,[4] kelp beds,[5] estuaries, and upwelling areas.[6] Deep oceans for instance, lacking photosynthetic input, have very low biomass productivity except for the hot-vent communities. Since plant life is impossible in the total darkness of the deep sea, food resources are consequently scarce. Most deep-sea food chains are nourished by organic debris that sediments down from surface waters where *phytoplankton* carry out photosynthesis. Concerning the abyss there are at least two processes by which materials are transferred to deeper layers: (1) by sinking of aggregates, in forms of both the so-called 'marine snow' and the carcasses of large pelagic animals; and (2) by the diurnal migration of large herbivores and carnivores (Angel 1997). Consequently, deep oceanic macrofauna feeding ultimately depends either on moving to upper levels to collect the food near the surface, waiting for food particles to sink, or for organisms to migrate at depth (COML 2001). Only 1 percent or less of surface productivity reaches the deep-ocean floor and consequently animal life is very scarce.

The major biological factors that determine living condition for marine biota are reproduction, predation, and competition. Regarding reproduction, marine fish are on average more fecund than freshwater species and their fecundity is directly proportional to the species' vulnerability to predation and inversely proportional to the protection given by their parents. Fecundity also appears to be inversely proportional to population size and this is why fish stocks which are severely depleted by over-fishing often show increased fecundity or reach sexual maturity earlier. Regarding predation and competition for space, ocean living creatures interact functionally through complex predator-prey relationships or food chains.

The basic marine food chain begins with the conversion of solar energy, which is primarily trapped by the anchored red, green and brown algae and a few plants in shallow waters and by *phytoplankton* producers in the open seas. Converting light and nutrients to food, marine *phytoplankton* manufactures most of the organic compounds required by the other marine biota. It has been estimated that about 300 billion tons of living plant matter are produced annually, which in turn support 5 billion tons of *zooplankton* biomass (O'Dor 2003). *Zooplankton* is made up of a wide variety of organisms ranging from tiny choanoflagellates filtering out bacteria to the extremely abundant crustacean copepods, appendicularians, and plantonic larvae from larger benthic organisms such as crabs, bivalves and many more. They primarily feed on *phytoplankton*. *Zooplankton* is in turn consumed by a wide variety of filter feeders, including shrimp, herring, anchovy and even the blue whale.

Predators such as shark, barracuda, moray eel, cod, salmon and killer whale represent the upper terminal link of the marine food chain. The shorter the chain, the more efficient is the production of terminal food chain animals in terms of weight and time to grow. This relation has a considerable importance in assessing sustainable exploitation of different species.

Regarding mobility fish conduct three basic types of migration: spawning, feeding, and over-wintering. Few species are non-migratory. Spawning migration ensures favourable physical and biotic conditions for developing eggs and larvae. Fish that live in the ocean most of their life but ascend freshwater streams to spawn are called *anadromous* (e.g. salmon). Fish that spend most of their life cycle in freshwater but descend to the ocean to spawn are called *catadomous* (e.g. eel). In addition, there are species that migrate from deeper water to shallow coastal areas in order to spawn (e.g. *Belone belone,* garfish). In cases of feeding migration, as updrafts or convection currents bring nutrients to the upper layers of temperate seas in winter, some commercially important species (*pelagic*) migrate into areas of nutritional abundance from considerable distances (e.g. tuna and swordfish). Some feeding migration tracks are vertical rather than horizontal, in rhythm with the movements of the plankton or in synchrony with the prey.

Over-wintering migration, which is conducted by only some species, is concerned with providing the most favourable wintering conditions for fish that are going to have a reduced metabolism and will therefore be less active. The distribution of certain marine fish species shows a high correlation with light intensity, day length, and water temperature. Even if fish distribution is also based on the other factors just mentioned, producing a computer-drawn isotherm chart is common practice in order to detect herds of commercial valuable species such as tuna and swordfish.

It is possible to classify species of fish according to the spatial distribution in the water column. *Demersal* species are those living closer to the seabed and are relatively non-migratory in character. They live and feed on continental shelf shallow waters and usually stick to the same area (called "fishing ground"). *Pelagic* species are wide-ranging, live in the open seas and in the upper area of the water column. They migrate over very large distances and consequently these living resources have a real 'global' nature. As a result, the areas where fish gather and are harvested may change according to temperature, light, season, and from one year to another, being even more marked for *pelagic* species. Some areas are therefore more productive for fishing and it must be kept in mind that only some species are economically important.

The size of *pelagic* fish schools depends on several parameters related to internal factors such as species, number of fish and psychological status, and to external factors such as current intensity, fish speed, and presence of predators. Acoustic surveys to assess *pelagic* stocks are expensive because of the high cost of R/Vs and, consequently, survey transects are widely spaced, data are sparse, and survey variance high. Therefore, mapping stock distribution is a difficult task (ICES 2002). All in all, when ocean living creatures are concerned, scientific uncertainty dominates the scene for lack of reliable data on quantities of fish, locations, genetic origins, interdependence of species, and relationships of fish species with marine ecosystems. Given living resources' high mobility, and density and vastness of the fluid they live in, oceanic biological research is difficult, time consuming, and involves high costs.

Traditional studies of marine organisms have involved their collection, preservation, and microscopic and taxonomic enumeration. Today the two main ways to approach these questions are through observational studies and modelling. To estimate fish spatial and temporal distribution, a complex function composed of different biological and environmental factors must be determined. Fisheries biological models have usually been correlated with just a single environmental factor. However, general models that correlate a large number of environmental parameters would be the ideal solution to better understand how many fish are in the sea and where they are located.

TECHNOLOGY TO STUDY THE BIOTA

One of the main issues in oceanic measurement technology is that the ocean is relatively transparent to acoustical energy in contrast to electromagnetic radiations. As addressed in the second chapter, the present technology for observing marine life is made up of different sophisticated remote sensing acoustic and optical applications. In addition, it involves modern classification techniques employing molecular technology. Molecular technology is a new field of research that allows for more accurate ways to identify organisms and name species. Molecular techniques use the unique genetic code found in individual organisms' DNA to identify and classify specimens.

Important advances in this field are the development of DNA probes and molecular bar coding. Bar coding, which is the use of a short DNA marker gene sequence to characterize species of organisms, is an efficient method to catalogue biological diversity. These technologies are currently being developed for use in combination with remote sampling platforms, thus enabling routine monitoring of environmental water samples for planktonic species.[7] Digitally acquired and processed data from remote sensing platforms marks all scientific advances of the last 30 years. However, the only way to directly count marine biota still consists of sampling systems. Fish must be either captured physically—with nets or video-controlled trap-systems—or detected with acoustical or optical instruments. This usually happens with the help of support platforms such as ships, moorings, and airplane over-flights.

To detect fish patterns of behaviour and spatial-temporal variability of fish abundance, the use of biological and chemical sensors based on remote platforms is a widely used method. Sensors must be calibrated and periodically adjusted according to rigorous standards. Many technicalities are still far from perfect; however, sufficiently accurate and reliable biological sensors—both optic and acoustic—are capable of recording appropriate variables. They have recently become available on the market at reasonable prices. Traditionally, sensors have been deployed at fixed depths. However, over the past decade several profiling instrument carriers have been developed and it is now possible to record data at different depths with the same device. Moreover, physical, meteorological and optical sensors have been integrated into autonomous platform systems such as surface and profiling floats and gliders. Sensor technology for marine biota measurement includes underwater LASER imaging (sheet LASER methods), acoustic tomography, long-range low-frequency back and forward scattering sensor systems, horizontal multi frequency SONAR and multi beam acoustics.

The technology that can penetrate the water is basically acoustical. Acoustic signalling offers the capability to assess the distribution of fish in a reasonably large mass of water (from 0.1 to 100 km^3). Modern acoustic devices employ multi beam echo sounders such as wideband and side scan SONAR and Acoustic Doppler Current Profiler (ADCP) to create images of large marine areas. Echo counting and echo-integration are basically methods for quantifying organisms. They have the capability to count and size individuals and, to some degree, they can also discriminate among species. Even a crab walking in the seabed can be detected with these methods. A challenge to the present use of technological devices in marine research is to integrate acoustic and optic technologies, thereby allowing a combination of high-frequency echo sounders or SONAR with video camera on the same underwater platform.

Conventional acoustic survey methods to assess stock biomass and distribution still use slow-moving R/Vs that send out high-frequency SONAR beams with highly localized line-transect methods. In this way, sound waves bounce only in a narrow column of water and, therefore, may under-sample fish population. The same methods are used in observing fish patterns of behaviour such as migration, spawning, predation, and competition. A new application of SONAR system for continental shelf areas has recently been developed in the US as a join project of MIT, Northeastern University, and the Naval Research Lab (Makris 2006). The new system uses SONAR at low frequency in order to make waves travel far distant, thus illuminating vast parts of the ocean and returning useful data over areas as wide as 10,000 km^2. Finally, in order to assess fish stock biomass, acoustic survey data is elaborated with classical statistical techniques—e.g. interpolation and integration methods—to calculate mean stock density and local stock density (biomass per unit area).

Acoustic and bio-optical techniques are also employed in tagging marine biota with tiny acoustic transmitters. Recently, there has been a proliferation of electronic tag types and systems for tracking fish (bio-probes). As a result, today researchers use a wide variety of tags to track fish patterns of movement.

There are fundamentally two basic types of tags: (A) Radio tag; and (B) Acoustic tag. Radio tags, which can only be used in water of very low salinity—as radio waves are attenuated very rapidly at sea—are only useful in freshwater.

Acoustic tags can be usefully employed at sea because sound is transmitted over long distances in salt water. There are two basic types of acoustic tags: transmitter tags; and transponder tags. The transmitter tag periodically transmits pulses of sound, allowing detection at remote locations. It can determine the presence of an animal and its direction. A wide variety of transmitter tags is available and inceases rapidly due to new developments for specific applications. Transmitter tags are larger than transponder tags and require an internal battery to power the transmitter and microchip. Tag sensors can record depth, swimming direction, speed and even heart rate of the fish.

Transponder tags emit sounds only when receiving an interrogation pulse from the SONAR, thereby saving battery power and extending the tag's life. Transponder tags consist of a small glass-encapsulated electromagnetic coil and a microchip. The tag is inserted into the body cavity of the fish either using a veterinary syringe or surgically. Transponder tags are commonly made in three different sizes, ranging from 11 to 28 mm in length and from 2.1 to 3.5 mm in diameter. They can be decoded with a portable hand-held reader and allow the position of a free-ranging fish to be fixed accurately relative to a R/V. According to CATAG (Concerted Action for Tagging of Fishes, Marine Research Institute, Iceland)[8] researchers have been using scanning SONAR equipped with acoustic transponder tags to track fish since the early 1970s. Until the 1990s, standard acoustic transponder tags of 8 mm in diameter and 51 mm in length, with a weight 8 g were used. However, they had been subsequently replaced with larger 'long-life' tags (16 mm diameter × 60 mm length, weight 16 g). The pace at which technology is reducing tags' size will contribute to a wider use of these devices in the future. Standard transponder tags have been used successfully to track plaice, cod, sole, dogfish, salmon, and eel. Data on time, location, pressure, light intensity, and temperature can be stored in the tag (archival tag) or transmitted to a compatible satellite-based receiver (pop-up tag). Unlike pop-up tags, archival tags must be recovered for data downloading. This implies the recapture of the tagged fish, which is usually generously rewarded (about USD 1,000) in order to increase recovery rates. Present recovery rate ranges between 30 and 40 percent. Pop-up tags are usually employed with large *pelagic* fish and mammals and data are available by satellite re-transmission (Block et al. 2005). Another recent type of tag is the Combined Acoustic/Radio Tag (CART), which is a hybrid tag combining components of both radio and acoustic tags. The use of genetic and chemical markers and tracers has also proved to have good potential in marine research (Alldredge et al. 1999).

Airborne measurement is another technique presently used to study marine biota.[9] One of the main advantages of airborne measurement is the speed of the aircraft, which ranges between 120 and 150 knots, allowing measurement of processes characterized by large spatial scale and short temporal scale. Airborne remote sensing technique is used to provide coverage of special features, especially where this technique is either more practical or more efficient than other methods. The major disadvantage of airborne measurement is in its limited depth penetration, which is only a few tens of metres.

There are two principal technologies employed in airborne measurements. The first is LIDAR. Airborne LIDAR, as already mentioned in chapter 2, has the same principle as RADAR and echo sounding, but instead of acoustic pulses it employs LASER pulses. Light from a LASER is expanded by a lens and directed to the surface by a pair of steering mirrors. The return light is passed through a polarizer into a telescope onto a detector. The detected signal is digitized and then recorded in a personal computer. Under very clean conditions, a standard LIDAR can penetrate down to about a maximum depth of 50 m. More sophisticated LIDAR surveys provide volume images or even horizontal images at fixed depth. The second airborne technology is multi- or hyper-spectral imaging, which employs ambient radiance. Depth penetration is not as great as LIDAR's and information contained in the images is therefore much less defined (Churnside 2000).

Another aspect of biological marine research concerns the possibility to observe phenomena continuously for an extended period of time. For these sustained types of marine observations there are now available platforms called *Eulerian Observatories*. Eulerian observatories are platforms equipped with multi-spectral sensor devices that remain at one geographic location for prolonged periods of time. Fixed to the seabed, these 'stations' can detect any part of the water column, the surface, and the seabed. They are widely used in meteorological, geophysical, and physical marine observations. Critical factors pertaining to Eulerian observatories are power requirements and data management, especially with regards to storage and transmission operations (POGO 2001). Fully equipped underwater laboratories are also employed for *in situ* research.

Even if adequate and satisfactory, approaches and techniques reviewed in this section are only suitable for local scale research. For larger scale assessment of marine living resources a global detection system is needed. Any project aiming at global observation must necessary rely on remote and autonomous platforms provided with larger area-coverage. A true global assessment of biomass abundance and distribution requires significant advances both in new technology and in better application of the existing ones, including enhanced acoustic, optic, genetic, and chemical sensors.

One of the most considerable improvements in global marine research is the application of satellite technology to study various atmospheric and oceanic phenomena. Ocean remote sensing through satellite-based sensors is presently the only tool for studying large-scale oceanic processes both in space and time. A global collaborative project involving more than 60 institutions from 15 countries and researchers from 70 nations began the Census of Marine Life in 2000 (COML Research Plan 2005). The world oceans have been divided in 6 realms and subsidiary zones according to their environmental characteristic, to the techniques, and to the difficulty of exploration. Different field projects cover all the realms listed. Animals as small as 25 g are expected to carry individually coded acoustic telemetry tags that will be recorded as they pass fixed monitoring lines.

The Ocean Bio-geographic Information System (OBIS) is the critical information component of the COML. Previously organized as a partnership of database sources, OBIS has developed into a global digital network of species database, including geographical references. For the period 2002-2006 about USD 18 million has been invested in OBIS. OBIS data is available on the Internet to a variety of users—i.e. all stakeholders concerned with marine life, including public and private managers of marine resources, ecosystem managers, fishery organizations, navies, scientists, educators and students.[10]

A special attention has been paid by COML to deep-sea *benthic*—i.e. bottom—habitats. Abyssal plains are the largest habitats of the world oceans. As the deep seas are expected to harbour vast numbers of species—most of which are still unknown—COML has a special section dedicated to the Census of the Diversity of Abyssal Marine Life (CeDAMar).[11] What it is known is that, in general, the abyssal ecosystems—i.e. below 4,000 m—have an extremely low productivity and hence biological rates are low (about 1/100 of the continental shelf ones) because of the scarce availability of food. Most of the food found in the deep seafloor environment is represented, as already mentioned, by scarce organic carbon that sinks down from the surface. Another scientific project regarding the depth of the oceans and led by the University of Hawaii (Eqpac Project) has recently discovered that across 3,000 km of the CCFZ, species accumulation shows 250 different species with large range of variability. Although there is a very rich biodiversity in the abyss, estimated in the order of millions, the largest proportion of the abyssal macrofauna is composed by small annelid worms (*nematodes*).

To detect and examine living resources' pattern of behaviour at deep sea is even more technologically challenging. A common deep-sea collecting technique for extraction of bottom samples is the use of a large sled, called *epibenthic sledge*, which is trawled by a ship. An epibenthic sledge consists of a rectangular steel frame with a mesh net (often more than one) attached to it. It is towed along the ocean floor and its weight scrapes into the benthos collecting any organism on the bottom and in the first few centimetres of sediment. It also collects organisms in the water column just above the benthos. A video camera is often attached to the net. Epibenthic sledges are used to collect relatively mobile benthic organisms. However, they cause injury to the seabed and to delicate organisms. Moreover, this technique is time-consuming and has the major shortcoming that specimens are exposed to the surrounding water temperature when the sledge is hauled at the surface. Thus, the animals might not survive the increased temperature and deform or become physically disrupted by the sledge, making identification very difficult.

Another sampling method for deep-sea examinations of benthic organisms is coring—basically plunging a big long hollow cylinder down into the seabed and closing the ends when dragging the sample to the surface. In this manner it is possible to obtain an undisturbed sample, where the only change will be in temperature as it is hauled up through the warmer waters near the surface.

FISHERIES

Even though certain groups of coastal populations have always relied on fishing for their living, from an economic perspective fishing became important for mankind about 500 years ago. The major marine living resources of commercial value have historically been finfish, shellfish (crustaceans, and molluscs) and mammals. Seafood provides high-quality, easily absorbable protein and a wide range of minerals and vitamins. It accounts for about 15 percent of the protein in the average human diet.[12] Worldwide, average seafood consumption per person has increased from 8 kg per year in 1950 to a peak of 17 kg in 1988. Since then, there has been a fall in consumption to the present 16 kg (FAO 2001a).

Although on a global scale fish consumption has doubled since the early 1970s, average per capita consumption of fish declined in developed countries from 24.3 kg per year in 1985 to 21.7 kg per year in 1997. However, developed countries' consumption of high-value species such as shrimps and salmon has increased[13] (Delgado et al. 2003).

The journal *Science News* estimated that in 1996 there were over 37,000 industrial trawlers and 12 million small fishing boats fishing the world's oceans, and approximately 200 fish stocks being harvested. Globally, an estimated 35 million people are directly engaged in fishing or fish farming as a full-time or part-time occupation. Fisheries post-harvest and service sectors provide a source of livelihood for tens of millions more people. Including fish processing, presently between 60 and 100 million workers derive their income from the fisheries sector (Delgado et al. 2003). Fishermen of the industrialized countries account for less than 10 percent of the world's total. Capture fisheries production in developed countries has grown since the late 1980s but in the last decades entered a crisis. About 85 percent of total decked fishing vessels and 50 percent of powered un-decked vessels are concentrated in Asia, while the top two capture fish countries, since 1992, are still China[14] and Peru.

As mentioned at the beginning of this chapter, marine living resources are unevenly distributed at sea and thus the productivity of different areas varies widely. This is reflected in the considerable variations of marine catches among regions. Coastal states' continental shelves account for more than 90 percent of the total marine catch, therefore, just 10 percent of the catch comes from high seas. Orange roughy is the most important commercial high sea species. The Northwest and Southeast Pacific are still the most productive fishing areas of the world. The top caught marine fish species are Anchovy (12 percent of the total), Alaska Pollock (3.5 percent), Skipjack tuna, Capelin, Atlantic herring, Japanese anchovy, and Mackerel (about 2 percent each). Tuna and tuna-like species account for 11 percent of the total value of landing for consumption. Giving the mobile nature of these living resources, national origin of fish is determined by the flag of the fishing boat that lands it, no matter where the fish has been physically caught.

According to FAO (2005), in 2003 the world production of food fish from marine capture and aquaculture was more than 130 million tons. Aquaculture—which accounted for more than 30 percent—has experienced a growing trend since the 1990s, while capture fisheries has seen a declining trend since 2000, partly because of the reduction in some fishmeal-producing countries, and partly because of several depleted or over-fished areas. Seventy-five percent of the total world catch is for direct human consumption, while the remainder is marketed as fish food, by-products for animal food such as fish oil and fertilizers for the chemical industry. Global trade of fishery products has been undergoing an upward trend in volume since the 1950s and the value of the international fish trade is growing accordingly. In 1998 about 40 percent of global fish output by value—corresponding to about 33 percent by weight—was traded across international borders. In 2002, with 50 million tons exchanged, the international fish trade market was an industry of about USD 58 billion in export value. Net fish exports from developing to developed countries was a business of USD 16.5 billion in 1999 (FAO 2001c).

The international demand for fish and fish products is expected to increase in the short and medium run, both in developed countries and in developing countries. However, the reasons for the expected growth differ, as there are different types of demand for fish. In developed countries, healthy food concern and high income stimulate high quality fish and shellfish consumption. In developing countries, the likely demographic growth and chip protein needs will arouse law quality small *pelagic* fish consumption.

Table 6.1 World Fisheries Production and Utilization (million tons)

Production	*1998*	*1999*	*2000*	*2001*	*2002*	*2003*
Inland						
Capture	8.1	8.5	8.7	8.7	8.7	9.0
Aquaculture	18.5	20.2	21.3	22.5	23.9	25.2
Total inland	26.6	28.7	30.0	31.2	32.6	34.2
Marine						
Capture	79.6	85.2	86.8	84.2	84.5	81.3
Aquaculture	12.0	13.3	14.2	15.2	15.9	16.7
Total marine	91.6	98.5	101.0	99.4	100.4	98.0
Total Capture	87.7	93.8	95.5	92.9	93.2	90.3
Total Aquaculture	30.6	33.4	35.5	37.8	39.8	41.9
Total World Fisheries	118.2	127.2	131.0	130.7	133.0	132.2
Utilization						
Human consumption	93.6	95.4	96.8	99.5	100.7	103.0
Non-food uses	24.6	31.8	34.2	31.1	32.2	29.2
Population (billions)	5.9	6.0	6.1	6.1	6.2	6.3
Per capita food fish supply (kg)	15.8	15.9	15.9	16.2	16.2	16.3

Source: Food and Agricultural Organization of the United Nations (2005) Progress in the Implementation of the Code of Conduct for Responsible Fisheries and Related International Plans of Action.

Since the mid-1980s to the late 1990s developed countries were the net exporters and developing countries net importers of fish. This situation has changed and now the three largest importing areas are Japan, the EU, and the United States. These areas—together with the other developed countries—account for more than 80 percent of the total import of fish products. On the other side, the top exporting countries are Thailand, the United States, Norway, China, and Denmark.

Developing countries account for more than 50 percent of the global fish exports and are projected to remain large net importers of low-value food fish and exporters of high-value finfish and shellfish. According to FAO's fisheries statistic, in 1996 about 50 percent of the fish product export volume was represented by "fresh, chilled or frozen fish" and about 20 percent was "fishmeal". "Crustaceans and molluscs" accounted for 15 percent of the market and "canned fish" for 8 percent (FAO 1997). In terms of value, fishmeal together with fish oil[15] account for 10 percent of the international market, while frozen fish and crustaceans have more than 70 percent of the share. The top-traded species are shrimp, tuna, salmon and *demersal* fish such as cod, hake, and pollock. Peru and Chile account for about 50 percent of the world fishmeal production. The production of fishmeal has historically been the second source of foreign currency earnings in Peru and the third in Chile, testifying the importance and dependence of these countries to ocean living resources.

There has been a large increase in the proportion of overexploited and depleted fish stocks from about 10 percent in the mid-1970s to about 25 percent in the early 2000s, figures varying widely among fishing regions and fish species (FAO 2005). Presently, about two-thirds of all fishing areas are subject to over-fishing. Hannesson (2004) distinguishes three degrees of over-fishing: (1) the most serious one, which occurs when fish stocks are exploited so intensively that they become extinct; (2) biological over-fishing—i.e. the depletion of fish stock to levels that support just a small sustainable catch; and (3) economic over-fishing, which is the case when labour, capital, and other factors of production used in fishing yield less than they should in economic terms.

Fish species particularly exposed to commercial overexploitation include those that congregate to spawn, such as cod and haddock, and those that migrate widely through many different EEZs, such as tuna and billfish. For each species, over-fishing causes a reduction in the size of the fish. Adding to natural mortality, over-fishing decreases the average life span of the species. Some species at risk such as blue fin tuna, blue whale and other wide-roaming *pelagic* animals used to congregate along oceanic fronts where cold and warm water masses collide because of the concentration of food.

Industrial fleets—which account for three-quarters of the global fish catch—consist of large-scale, capital-intensive vessels that practice distant water fishing and use advanced technology to locate, catch, and process fish. Satellite maps of sea-surface temperature and spotting planes are common in order to track and harvest precious fish stocks. A blue fin tuna can grow up to more than 600 kg and can be sold for up to USD 70,000 in the Japanese market (Malakoff 2004). Even though recently the trend started to decrease, world marine fisheries boomed from a catch of 20 million tons of wild fish in 1950 to a catch of more than 95 million tons in 2002. However, over 25 percent of the total world catch is not saleable and is thrown back dead into the sea. Other shameful wastes occur when only a small portion of the fish is used, such as the shark's fin. FAO alerted several times that 11 of the world's 15 major fishing grounds are depleted. Stock collapse as those experienced with Atlanto-Scandian Herring, Pacific Anchovy, and North Sea Cod seem to suggest that managing marine living resources is a difficult task both at a national and international level.[16] Recent scientific research indicates that fish populations are less resilient than once imagined and that the recovery rate of depleted populations is much slower than expected.

Capture fisheries

The fisheries industry is one of the most risky in the world because successful business depends on forces of nature such as weather and abundance of fish, over which fishermen have no control. The international fisheries market is very competitive, each fishing vessel racing to exploit common open-access resources before someone else does it. Instead of investing in the long-term renewability of fish population (by leaving young fish in the water) fishermen are compelled to invest only in new, sophisticated equipment and fuel, thus increasing their chance of catching fish before their competitors do. Even though there are developing countries where the common fishing technology among the artisan fishing community is still represented by a dinghy and a fishing rope (and catching may be vital in order to survive), fishing technology has improved greatly in recent years and very sophisticated devices are commonly used to track fish stocks.

Advances in fishing technology have boosted heavily fishing capacity. Refrigerated processing vessels have enabled fishermen to exploit distant oceans efficiently. Distant-water fishing vessels are real 'fishing factories'. They are usually state-of-the-art technologically equipped, including quick-freezing and automated filleting machines. New factory trawlers are hundreds of feet long and are capable of processing over 600 metric tons of fish per day. They can access fishing grounds located in the EEZ of other countries by purchasing access fishing rights. Many coastal developing countries sell an access right to fish in their waters to foreign governments or companies, mainly because they lack the capacity to fully exploit their own living resources. Consequently, fishing agreements between countries is a common practice worldwide.

The classic economic theory of capture fishery[17] (Gordon 1954) assumes that yield is proportional to: (A) the amount of effort put into fishing; and

(B) the level of the fish stock. Gordon defines the optimal utilization of a fish stock as the one that maximizes the net economic yield—i.e. the difference between total cost and total receipts. The total cost curve is a linear function of the fishing effort or intensity, while the production function increases at a diminishing rate as the fishing effort expands. Moreover, this theory assumes that the amount of fishing effort is not affected by the cost of fishing gears. Therefore, marginal costs (MC) and average costs (AC) are identical and constant, and can be represented with a straight line.

As in microeconomic theory, the optimal level of exploitation (the one that provides the maximum net economic yield) is the point where MC and marginal productivity (MP) intersect (point D in Figure 6.1). The maximum sustainable yield (MSY)—which is the largest amount of fish that can be harvested annually from a fish stock without depleting it—is the point where MP of the fishing effort is zero (point Z in Figure 6.1).

Therefore, the fishing effort that corresponds to the economic optimum (point D) is less than that of MSY (point Z). However, as fish are not privately owned, there is competition among fishermen on the same fish stock, which in turns dissipates the associated rent (rectangle ABCD in Figure 6.1). This basic inefficiency in capture fishery production explains why fishery has been traditionally put under some forms of spontaneous social control (cooperatives, associations, etc), that in some fashion turns the open-access common nature of these resources into a property right regime. Subsidies policy brings in even more inefficiency in capture fishery production.

Bio-economic models of capture fisheries tend to integrate economics with biology. They have been largely employed in capture fishery research. In these models the efficient management of fisheries is a function of biological parameters. The most common biological parameter used in bio-economic modelling is the fish stock natural growth function. There is

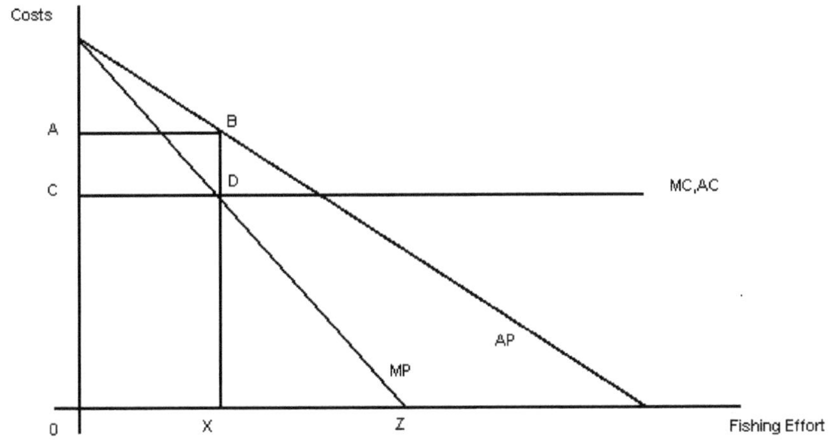

Figure 6.1 Economic Model: Fishery Market Allocation.

consensus among researchers that one of the strongest points of biomass dynamic modelling is its simplicity. Generally, bio-economic models of capture fisheries try to assess the amount of return from fishing in a variety of scenarios with different economic conditions. Environmental conditions and fish population dynamics are usually assumed to be constant, mainly because the specific mechanisms are generally unknown. However, it is known that fish population dynamics involve interrelation effects of biotic interactions, such as predation and competition; and environmental variability such as shifts in temperature, salinity, and oxygen content.

The natural growth rate of the population of each species cannot be represented by a decreasing function, nor by an increasing function, otherwise the population will approach either extinction or infinite growth. Population natural growth is usually described by a sigmoid curve or logistic function. This is a dome shaped curve that might be symmetric or non-symmetric (Wang and Wang 2005).

Figure 6.2 represents the fish population natural growth rate. The horizontal axis measures the population size and the vertical axis its growth rate. There is only one peak in the growth rate that corresponds to stock size S*. S1 represents the stock extinction, while S2 is the biological equilibrium size that corresponds to the carrying capacity of the environment—i.e. the point where the population would persist in absence of disturbance. Due to natural dynamics of self-regulation of species, such as in/out migration and mortality and birth rates, we have two different phenomena:

(A) when the population size is between S1 and S*, the growth rate increases as the population size increases. A greater size always implies a greater net production;
(B) when the population size is between S* and S2, the growth rate decreases as the population size increases. In this case, a greater size always implies less net production.

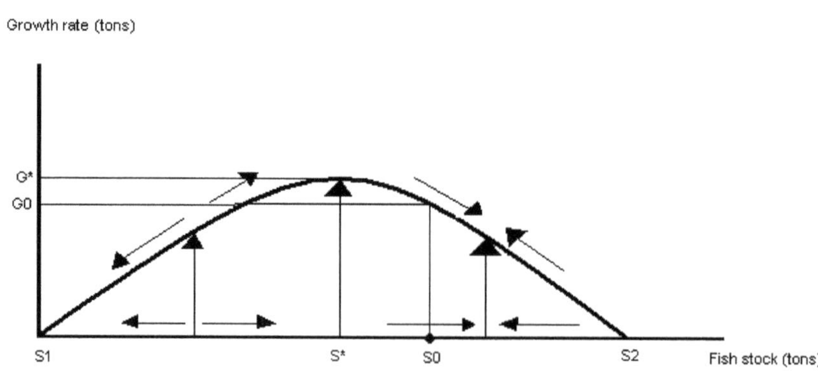

Figure 6.2 Biologic Model: Growth Rate of Fish Population, based on Wang and Wang (2005).

Once human disturbance is introduced (such as fishing), stock dynamics depend on both the current status of its size and the strength of the disturbance (measured by fishing effort and technology). If at the moment of disturbance the status of the stock is between S* and S2, the stock tends to adapt and converge to the disturbance. If at the moment of disturbance the size of the stock is between S1 and S*, there are two different results according to the strength of the disturbance. If the disturbance is greater than the net production of the current stock, the population will diverge and approach collapse. On the other hand, if the disturbance is less than the net production, the stock tends to adapt and converge to the disturbance. An aggressive harvesting strategy can therefore accelerate the depletion of a fish stock if employed in times of poor net production. Therefore, the harvesting regime can intensify or alleviate the consequences of environmental changes to population dynamics. The sustainable harvest rate is the one that equals the growth rate of the stock. In Figure 6.2, S* is the population size with maximum growth and therefore corresponds to MSY (point G*). However, a sustainable yield can be determined for every stock size by drawing a horizontal line from the known population size to the vertical axis (e. g. Stock = S0 ⇒ Harvest = G0).

A classic starting point to determine the efficient maximum sustainable yield (EMSY) in capture fishery economics is the bio-economic model developed by Schaefer in 1957. This model was originally developed to calculate the equilibrium yield under given conditions. Variables such as water temperature and age structure of populations are kept constant. The biological growth process is described by a logistic function similar to that represented in Figure 6.2. Schaefer's model determines a relationship between growth and size of the fish population (Tietenberg 2006). Under stable environmental conditions, fish stock dynamics are purely dependent on the current status of its size. The rate of change of the fish stock over time is composed of an inflow and two outflows. Increase in population biomass is a single inflow due to both growth and addition of new fish (recruitment).[18] Therefore, addition to biomass is a fixed fraction of the current biomass of the stock. Decrease in population biomass is composed by two outflows: (1) natural death rate; and (2) the harvesting rate.

In equilibrium condition the inflow to the stock must equal the sum of the two outflows. To determine EMSY, the harvesting costs must be included in the model. In static conditions the total costs (TC) of harvesting the fish stock are assumed to depend linearly on the fishing effort and can be measured by vessel years or hours of fishing.[19] The revenue function—assuming that the price of the fish remains constant and the demand curve is perfectly elastic—has the same dome shape of the graph in Figure 6.2. The fish stock size is a negative function of the level of effort: the more fishing effort the less fish population. Therefore, each and every effort level corresponds to a population level in Figure 6.2. Net benefits are the difference between total revenue from sales of landings—i.e. the price of the fish multiplied by the

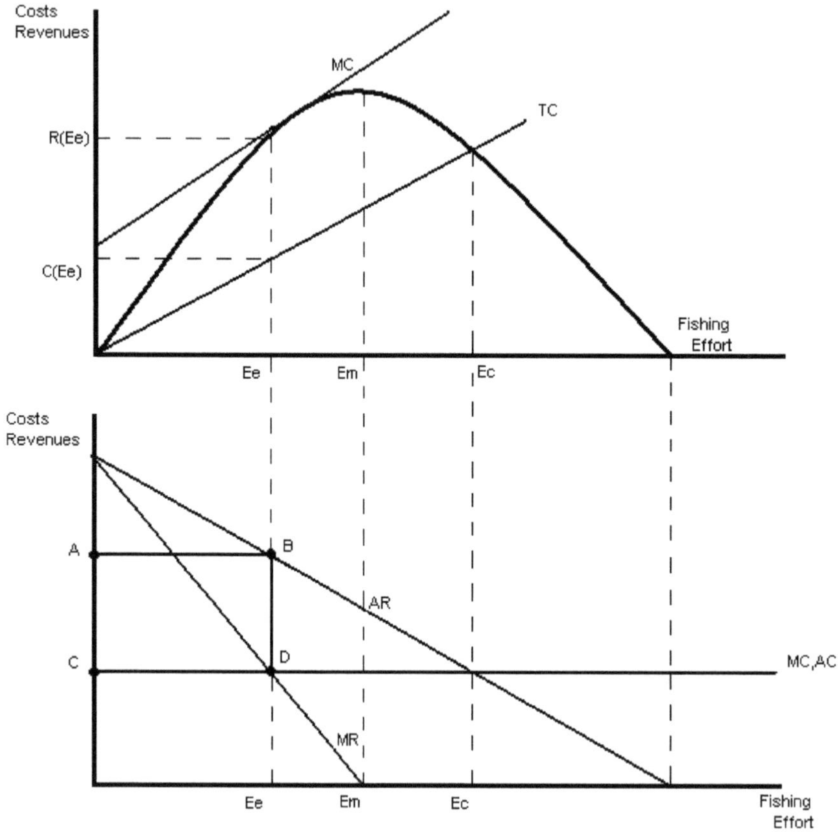

Figure 6.3 Bio-economic Model: Maximum Sustainable Yield, based on Tietenberg, Environmental & Natural Resource Economics, Figure 13.3 "Market Allocation in a Fishery", p. 294, © 2006, 2003 Pearson Education, Inc. Reproduced by permission of Pearson Education, Inc. All rights reserved.

quantity caught—and TC—i.e. the marginal cost of the fishing effort multiplied by the units of effort expended. In Figure 6.3, point Em corresponds to MSY (point S* in Figure 6.2), while point Ee corresponds to EMSY. At EMSY, MC equals MR, and the efficient level of fishing effort is optimized. Hence, MSY does not coincide with EMSY because the maximization of a physical quantity does not necessarily maximize economic benefits.

In the dynamic model of EMSY, positive discount rates are introduced and EMSY, in static conditions, is just a particular case of dynamic (or inter-temporal) EMSY in which the discount rate is zero. When positive discount rates are introduced in the model, the efficient level of effort is higher, therefore, the catch and the net benefit increase accordingly. However, since the increased level of catch exceeds the sustainable yield for the stock size, fish stock population would be reduced, thus reducing future catch levels. With an infinite discount rate, a lower equilibrium level would be eventually obtained in Ec, the point where net benefits equal zero. The same result is obtained when fisheries have unlimited access. The equilibrium point of open-access common property capture fishery would be the point where total revenues equal total cost, fishing effort is expanded to Ec and profits equal zero. With private property fishery—the profit maximizing market case—marginal cost would equal marginal revenue and the rent (the rectangle ABCD in figure 6.1) would be maintained (point Ee). In summary:

1. the open-access capture fishery solution produces the lowest level of catch associated with the highest level of effort;
2. the MSY solution gives the highest level of catch; and
3. the EMSY optimal market solution gives the lowest level of effort coupled with the highest level of profit.

What is observable in practice is that the catch trend for all commercial fish stocks usually begins at a low level and then increases over time because the demand for that fish species increases. Higher prices reward fishermen for investing in fishing equipment to catch more and as a result the fishing technology improves. The increase in catch continues—since technology improves and more vessels enter the fishery—until the stock cannot sustain the fishing pressure anymore. Then fish abundance starts to decline and catches start to decrease accordingly, eventually falling to zero. Fish stocks are usually exploited from more abundant to less abundant species, from easily caught to less easily caught species and from high valued to less valued species.

With regards to extinction of species, usually a fishery collapse economically before species become endangered. Actually not many fish stocks seem to have gone extinct. The main reason is that catch per boat falls as the fish stock is driven down. As fish become fewer, they are more difficult to find and catch. Normally fishermen enter a fishery because profits seem attractive. Profits are linked to the level of catch per unit of fishing effort (CPUE). If CPUE level is high, additional effort will be applied to catch more fish when they are still abundant. As more fishermen enter the fishery, fish stock biomass starts to drop and CPUE declines. When CPUE level does not cover the total costs, profits will turn to losses. Once this new condition has been reached there will be no incentive for new fishermen to enter the fishery. The marginal fisherman will leave the market and the remaining fishermen will intensify their effort and attempt to improve the efficiency of their fishing gear. A new equilibrium will be reached when the catch is equal to the surplus growth.

Culture fisheries

The cultivation of marine species is an ancient practice, traces of which can be found in China as far as 1000 BC. At that time, carps were already farmed in rice fields. Culture fisheries or aquaculture, is considered the agriculture of the seas, since many of the same land farming management techniques are employed. Culture fisheries involve growing selected species with the commercial purpose of harvesting the production in order to sell it. The practice of culturing only a single species is called monoculture. In some cases several selected species are grown in the same controlled environment. This practice is called poly-culture. There is a broad variety of systems and practices in aquaculture, ranging from backyard household ponds to large-scale industrial companies. Hundreds of varieties of finfish, molluscs, crustaceans and aquatic plants are cultured all over the world.

Given the widespread depletion of wild fish stocks, aquaculture can be a means of supplying high-quality food to a growing human population in a sustainable way. In 1984 aquaculture represented just 8 percent of the world fishmeal consumption. During the 1990s and the early 2000s, aquaculture production developed at an annual average rate of nearly 10 percent. Today aquaculture accounts for over 30 percent of the seafood production (FAO 2001b). From 1985 to 1997 aquaculture was responsible for 71 percent of the total growth of seafood production by weight (Delgado et al. 2003). The high rate of growth of the aquaculture industry (see Figure 6.4) is the consequence of structural changes that occurred in the capture fishery market, where the increase in the demand for seafood is facing a supply characterized by a downward trend. Moreover, with the development of aquaculture production the availability of high demand species is increasing and the market supply is more stable. Given the increasing demand for

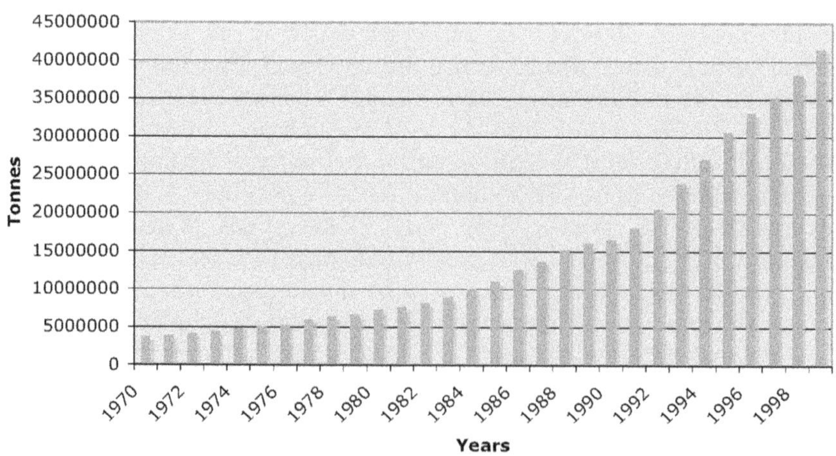

Figure 6.4 Global Aquaculture production trends by major species groups 1970–1999, based on Group of Experts on the Scientific Aspects of Marine Protection—Food and Agricultural Organization of the United Nations (2001) Planning and Management for Sustainable Coastal Aquaculture Development.

seafood and the reduction in wild fish stocks, an even faster growth of this industry is expected in the years to come.

According to FAO, about 42 million metric tons (mmt) of cultured fish were produced worldwide in 2003 corresponding to a value of USD 53 billion (FAO 2004). Aquaculture production in developing countries rose from about 2 mmt in 1973 to over 25 mmt in 1997 and this impressive trend is expected to continue (Delgado et al. 2003). About 60 percent of the total aquaculture production of seafood comes from freshwater. In 2002 Asian countries accounted for more than 90 percent of the global production. China alone accounts for more than 60 percent of the total aquaculture production. More than 50 percent of the global aquaculture production comes from coastal areas marine farming (mainly seaweed and molluscs).

There are four main farming techniques employed in aquaculture production: pond, cage, raceway and re-circulating systems. According to the level of capital input and to the corresponding production output three grades of farming operations can be distinguished: (1) Extensive; (2) Semi-intensive, and (3) Intensive.

Extensive aquaculture production is a low-density traditional system where small capital inputs and little effort are employed. The reason is often lack of capital. Semi-intensive aquaculture operations imply more capital input, such as compound aquafeeds and water control technology. Yield is usually higher than that of extensive aquaculture farming. Intensive aquaculture methods have high capital input and consequently production and yield are higher than that of the other methods. Intensive farm methods rely on compound feeds, seeds from hatcheries, chemicals such as pesticides, antibiotics and other capital input such as aerators, water control technology, etc.

Modern marine fish farming is basically the placement at sea of large metal or mesh net pens to breed finfish. This technology was pioneered in Norway in the 1960s and later expanded to other countries such as the UK, Ireland, Canada, the United States and Chile (OECD 2004). Global aquaculture has focused on species that command a relatively high price. Presently Norway is the world's largest exporter of Atlantic salmon. Shrimp farming has been developing very rapidly in recent years and large-scale, intensive aquaculture productions for export markets occur especially in South-East Asia, where this phenomenon recently started to raise environmental concern. From 7 percent in 1970, this industry now has a share of about 50 percent of the global shrimp production, with a market value of almost USD 10 billion. The top producing countries are Thailand, Indonesia, Ecuador, India, China and Vietnam (Deere 2001).

The most vulnerable points of aquaculture farming are organism diseases, algae and jellyfish, physical damages, and escaping fish. High stocking densities, poor water quality and poor seed quality are the principal causes of disease outbreaks, which is a major threat to production. Disease control may be partly addressed by management techniques such as water control technology, antibiotics, rotation of culture species and lower stocking densities. Marine biotechnology may be usefully applied for improving safety in aquaculture production. Aquaculture may be improved by enhancing species characteristics with useful features such as fast growth, resistance to pathogens, and high temperature tolerance (Genetically Modified Organisms, GMOs). However, up till now transgenic technology is not widely used because of customers concern regarding GMOs consumption. Although no genetically modified fish have yet been approved for commercialization, some species have been modified for enhanced growth and farmed all around the world (Delgado et al. 2003). Other improving features made possible by biotechnology are related to the development of optimal feeds and feeding, increase in the health of cultured populations, and to resistance to diseases, as well as minimal environmental impact.

Environmental impact of aquaculture

Usually short-term incentives of immediate profits overpower long-term costs of sustainable techniques. Therefore, aquaculture management decisions have usually paid scarce attention to environmental issues, especially in developing countries.[20] There are three main environmental threats related to aquaculture production, the magnitude of which is higher for intensive farming methods:

1. the impact of wastes and chemicals;
2. the impact resulting from interactions between farmed fish and wild species; and
3. the impact on wetlands and coastal areas.

The dimensional extent of the effects of wastes and chemicals can be near-field[21] and far-field. Near-field effects are bounded by the physical limits of particulate waste dispersal and sedimentation from individual cages or farms. Far-field effects extend to wider areas or sub-ecosystems. Fish farms are a source of suspended and dissolved organic matter originating as fish faeces, excess fish feed and net-cleaning wastes. Organic matter is transported from fish cages to the surrounding marine environment by the action of all types of water movements in the immediate vicinity of the cages. As this material moves, it spreads out and its concentration decreases, both by dilution and sedimentation processes. The impact of wastes includes therefore both nutrient and organic matter, and may provoke eutrophication,[22] sedimentation, and other adverse effects on the marine food chain.

Long-term eutrophication of coastal and estuarine waters result from additions of both dissolved inorganic and organic nutrients. Both these factors increase biological oxygen demand at the bottom. In this process the dissolved inorganic nutrients released by finfish culture stimulate *phy-*

toplankton growth, leading to an overall increase in *phytoplankton* biomass production. This surplus production will eventually sink down to the bottom, greatly increasing the amount of organic matter in the upper layer of the bottom sediment. The organic matter will, in turn, be degraded primarily by white sulphur bacteria, leading to a massive consumption of oxygen in the upper sediment. This increases biological oxygen demand from oxygen-consuming marine organisms. Unconsumed feed pellets and fish faeces contribute to increase local concentration of suspended and sedimented particulate matter.[23] Water depth and current velocity are critical factors in determining the pattern of sedimentation around cage sites and therefore the impact on benthic fauna is different according to the farm sites. Intensive bivalve aquaculture, for instance, has the capacity to alter matter and energy flows at coastal ecosystem scale. Mussel culture operations can significantly reduce *phytoplankton* biomass through grazing.

Chemicals used in aquaculture farming usually include pesticides, drugs and anti-foulants. There has been a massive scientific debate regarding the environmental consequence of chemical usage in aquaculture. Food additives include antioxidants (preservatives) and carotenoid pigments (flesh colouring). However, these chemicals are now considered unlikely to cause any effect to the environment.

The impact of interactions between farmed fish and wild species includes both disease transfer and genetic and ecological interactions among species. The main threat seems to be the possibility that escaped farmed fish may interbreed with their wild counterparts, thus modifying the natural genetic pools of wild species. Research has showed cases of successfully reproducing escaped salmons.[24] Although there is still no assessment about the consequences (Delgado et al. 2003), harmful effects may certainly be provoked by interbreeding of wild species with genetically modified ones. Another issue regarding species interactions is related to fish feeding in aquaculture farming. The extensive use of fishmeal and fish oil as feed input in carnivorous and omnivorous fish has increased the demand for this commodity. Fishmeal and fish oil are produced from small pelagic fish such as Anchovy. Therefore, if on the one hand aquaculture development is believed to ease pressure on overburdened and overexploited capture fisheries because of the substitution between capture and farmed fish, on the other hand, there is concern that fish farming will place increased pressure on wild pelagic stocks.

Another environmental problem caused by the fast and considerable aquaculture development is coastal and marine habitat disturbances. The main threat concerns the replacement of valuable unique marine ecosystems with farming areas. Destruction of mangrove[25] forests to grow shrimps, especially in South-East Asia (Indonesia and Philippines) and in Ecuador, has been of major concern for environmentalists. Between 1976 and 1990 Ecuadorian shrimp farmers cleared more than 1,000 km^2 of mangrove to build shrimp ponds. The roots of some mangrove species form props to the trunks of the trees creating surfaces on which algae, barnacles (crustaceans) and other important organisms of the marine food chain settle. Therefore, mangroves provide a nursery habitat for numerous species of fish and shellfish. Cutting down mangroves for shrimp farming harms wild species and, consequently, traditional fishermen who rely on mangrove fishing habitat for their catch. It has been estimated (Delgado et al. 2003) that for every hectare of mangroves transformed into fish ponds, more than 100 kg of fish within mangrove ecosystems is lost and that each kilo of shrimps produced in former mangrove areas results in a loss of over 400 g of wild fish.

Sustainable aquaculture

To achieve sustainable aquaculture development due attention should be paid to water management, effluent control, disease control, and minimization of land use (Delgado et al. 2003). One way to attain sustainable aquaculture farming systems is to create a small ecosystem with a variety of species being harvested. As already mentioned, this practice is called poly-culture.

Poly-culture is an integrated system of aquaculture farming that includes different species of fish in which the waste of one species provides food for another. In this way water filtration, removal of organic matter, and increased production can be achieved at the same time. Poly-culture can be a means to overcome environmental problems while aiding in parallel productions of other useful products such as bivalves, seaweeds, and high-protein *phytoplankton* biomass. Ideally, species populations in poly-culture farming should be arranged so that nutrients introduced into the system are recycled among organisms, thus reducing nutrient waste and pollution. One such example of poly-culture is growing seaweeds or oysters along with fish species. Seaweeds and oysters act as a filtration system utilizing fish wastes, which would otherwise be released into the surrounding environment resulting in nutrient pollution. Natural mechanisms increase the efficiency of the system while minimizing the impact of the farm on the environment.

Regarding regulation of aquaculture development, even though many countries have some form of legal framework in place to regulate this issue,[26] adequate procedures to undertake environmental assessments and monitor aquaculture operations are still lacking. In 1976 the FAO held a meeting in Kyoto, Japan, which was the first major conference on aquaculture and adopted a declaration. In 2000 FAO and NACA (Network of Aquaculture Centers in Asia-Pacific) organized a conference on aquaculture development, which discussion resulted in the Bangkok Declaration and Strategy (NACA/FAO 2000).[27] This document encourages states and the industry to adopt a number of measures to reduce the environmental impact of aquaculture emphasizing long-term sustainability and regulation of production standards. The Bangkok Declaration focuses on key elements of aquaculture production and suggests investing in education and training and in research and development, to improve environmental sustainability, to apply biotechnology, genetics, and innovation in aquaculture, and to manage aquatic animal health. It declares: "...aquaculture policies and regulations should promote practical and economically viable farming and management practices that are environmentally responsible and socially acceptable".

FISHERIES MANAGEMENT

Fish stocks are a type of renewable resources sometimes called "interactive resources" (Tietenberg 2006). Being interactive means that the size of the fish stock is determined jointly by biological considerations and by actions taken by humans. Since human actions determine the present harvesting rate over time provoking inter-temporal effects, in order to conserve fish stocks, an efficient sustainable level of harvest must be determined. A complex mixture of biological, social, economic, and environmental information affect the decision-making process of fisheries management.

The basic governance problem in marine living resource management is to provide sufficient means for meeting three main tasks. The first is gathering a reasonable scientific knowledge of fish stocks' size and dynamics in order to assess whether and how harvesting should be conducted—i.e. to define the 'optimal fishing' level and, consequently, the Total Allowable Catches (TAC). Long-term management of commercial fisheries requires an even better and thorough knowledge of the resources and their dynamics.

Traditionally, estimates of fish stock populations have been based on fishermen's catch or on data from trawl surveys collected by government agencies. Today the three main sources to estimate the size of commercial fish populations are: (1) data from landing at ports (total international landings and market sampling),[28] (2) data from fishermen (logbooks,[29] questionnaires and sampling of discards), and (3) R/Vs surveys. R/Vs survey data is cross-checked with the two other sources of information to confirm the reliability of the findings. Researchers use the same fishing gears that fishermen use—i.e. bottom and beam trawls.

A wide variety of sophisticated acoustic assessment methods, such as side scan or multi beam SONAR, are available to assess fish stocks. However, in order to compare data over the years, the same fishing detecting methods should be used without introducing novel techniques. Another way to assess commercial fish stocks is an indirect method that consists of counting eggs and larvae captured with very fine mesh nets. Since the number of eggs and larvae should be proportional to the adult fish that produced them, researchers can infer the adult fish population. Once data on fish eggs and larvae is collected, researchers use mathematical and statistical models to obtain fish population estimates and forecasts.[30]

The second task of fisheries management is to consider the social and economic implications of the biological advice. This should be done through the adoption of appropriate regulatory measures in order to govern catching activities, such as for instance determining the extent of the fishing effort or, alternatively, the fishing catch. Regulatory measures should ensure long-term conservation of fish stocks (the long-run equilibrium biomass or MSY) and allocate living resources in an agreed-upon way based on the regenerative capacity of the fisheries. In theory, optimal depletion rates for fish stocks should first be assessed. Then, in the framework of the harvest strategies and tactics selected, MSY defined and the relative TAC respected. Harvesting strategies can be formulated with the active participation of fishermen and processors.

There is a whole set of strategies and different related outcomes available to policymakers in fisheries management. The simplest one regards fish stocks and aims to maintain constant either the stock's size or the exploitation rate. Legal measures can be implemented to regulate:

1. the fishing effort, by determining the number of vessels allowed in the fishery, the vessel capacity, the fishing intensity and the permissible time or fishing area per vessel;
2. the permissible catch, by determining TAC within a specific fishing area, quotas for individual fishermen or for fishing cooperatives, or even legal limits on the size of the fish that can be caught; and
3. limitations on where, when and how fishermen may fish, by setting specifications on fishing tackles such as lines, nets and other fishing technologies, and by closing areas of gathering and spawning seasons.

The third task in fisheries management is enforcement. This problem is twofold, as both (1) an effective system to promote compliance with the measures adopted among the users of the resources at stake and (2) surveillance are needed. In other words, this requires that results from marine biological investigations should feed into decision making processes on strategic regulations, carrying out fish harvesting in a sustainable manner. Due to the lack of data available to assess sustainable or optimum yields of target species, the precautionary approach[31] has been introduced in the fisheries system at all levels by FAO's Agreements and Codes in order to conserve and manage living stocks. However, the main issue in fishery management is the difficulty to enforce policies because of the open access nature of the resources.

Today, because of the assessed exploitation[32] of the genetic diversity of the ocean and the erroneous conclusion of some fisheries' research, priority in marine living resources management is resource sustainability and better understanding of ecosystems. Fisheries research was previously focused on catching target fish more efficiently. Interest and focus have expanded to encompass a more highly integrated management of human activities in the oceans. All activities should be managed in a wider framework of ecosystem considerations.[33] Hence, today the primary role of research in fishery management is to provide managers with the information needed to protect and conserve the marine ecosystem.

The ecosystem approach represents a more holistic and interdisciplinary strategy, in which isolated problems are integrated in order to acquire a better understanding of the complex ecosystem dynamics.[34] A fish population has an important role in the ecosystem as predator, prey, and as a contributor to genetic diversity. In the context of advanced fisheries management approach (from fish stocks to ecosystems), acoustic technology is essential, since it is the only tool providing simultaneous observations on organism interactions and substratum characteristics (ICES 2002).

System ecology is the science that studies the interactions of human population with the environment. One of the major concepts in this discipline is the optimization of ecosystem exploitation through sustainable management. During the last decades, there has been interest in fish stock assessment research in order to provide scientific foundation in determining fishable biomass and consequent harvesting quotas. Bio-economic models usually assume that changes in the characteristics of the fish population are expected to be due only to changes in the pattern of the fishery. Studies on fisheries have shown that these assumptions were not completely valid. The adaptation of fish stocks to changes in environmental condition can modify the fundamental characteristics of the parameters considered in conventional stock models. Population ecology is the branch of ecology that studies structure and dynamics of populations, providing models that can be used to gather information on fish stocks. It is one of the most formalized areas in biology.

In studying the dynamics of animal population, May (1989) found that many populations usually fluctuate. Populations tend to increase after having dropped to low density, however, after having reached high density they tend to decrease again. All in all, population values seem to be random, show high sensitivity to initial conditions and are thus driven by chaotic dynamics. In addition, a population usually interacts with other species, which, in turn, interact with others and such a web of interactions makes chaotic dynamics even more likely to appear.

Understanding population dynamics is a formidable task and, consequently, long-term prediction becomes almost impossible. Such understanding has practical applications in trying to predict the likely effects of both natural and man-made changes, such as when the climate pattern alters or when a population is harvested. May's approach of distinguishing between chaos and random noise in population biology is still in its infancy. However, population ecologists were among the first scientists to realize that mathematical chaos may drastically change our understanding of how the dynamics of natural phenomena work. After intensive theoretical and empirical investigation over the past two decades, there is still no widely accepted example of chaos in field populations. Such methods only indicate that there are some nonlinear mechanisms at operation without stating what the mechanisms are.

Baseline information on natural phenomena is difficult to gather and human impact on nature even more difficult to assess. Sometimes, in the absence of data about fish stock assessment, fishable biomass and consequently harvesting quotas are fixed at precautionary levels. However, precautionary levels of fish biomass can either underestimate or overestimate the real stocks' size and thus the availability of resources.

One of the most common approaches to establish regulation in the fishery sector is the 'single-species' approach. Protected species and catching limits are determined according to species, often without taking into account ecological interactions such as competition or predation. Usually the best available information managers possess does not include assessments concerning natural long-term trends and variability of fish population parameters, nor ecological interactions with other species. Consequently, a high level of uncertainty characterizes decision-making in fishery management. Unreported and illegal catches further decrease the reliability of fishery data, which is the basis for policy decisions.

There has been a gradual evolution in fisheries management from the concept of harvesting an independent single species towards ecosystem management under conditions of uncertainty. Due consideration to both the multi-species approach and at the ecosystem level is therefore paid, to some extent, in the assessment and management of fisheries resources. Many countries lack adequate technical and financial means to sustain activities such as scientific research, resources assessments, and effective monitoring and surveillance to enforce their policies. However, control and limitation of quotas are usually required anyway in order to try to preserve fish stocks.

Regulations and catches are traditionally controlled through the use of logbooks, with landings also checked in port and by sea patrols. Monitoring, control, and surveillance activities are now performed with the help of technologies such as satellite tracking and telemetry. This makes it easier for managers of fisheries to detect illegal or misreported catches.

One of the latest technological advancement in regulation is vessels monitoring system (VMS), a technique that employs satellites to track vessels movements, thereby enhancing the enforcement and deterrence of regulations. Given the limited resources available for patrolling vast areas of ocean, remote monitoring is an extremely valuable and cost effective management tool. VMS is already used in several countries. However, controls on fisheries still depend heavily on voluntary reporting, and thus on cooperation and trust between managers and fishermen. Amidst regulations, the exclusive use right is an interesting method for achieving an efficient exploitation of fish stocks. The Individual Transferable catch Quota (ITQ) is the most widely applied form of exclusive use right. ITQ is the right to catch a specific quantity of fish from a given stock within a given period of time. Under ITQ programmes, fishermen are allocated fixed quantities or percentage shares of the overall quota. Among the different criteria to allocate quotas, the rule based on catch history is the most used one, even if different criteria can be used simultaneously, such as for instance the vessel length and the catch history.

Quota holders have incentives to maximize the value of the catch they are allowed to fish and to minimize the related operating costs. Long-term tenure of quotas is essential both for investing optimally in fishing vessels and for preserving the productivity of fish stocks. In addition, since fishing is a highly risky business, one of the arguments in favour of fishing rights is that fishing would become less uncertain.

More often than not, half of the fishermen that participate in the initial quotas distribution sell them, and the remaining fishermen begin to fish when they are most likely to receive a higher price per catch on the market. However, there are problems related to the use of ITQ, such as that many fisheries catch more than one species, landing limits are often disregarded, and underreporting is still a widespread practice among fishermen.

Since UNCLOS and its Implementation Agreements relating to the Conservation and Management of Straddling Fish Stocks and Highly Migratory Fish Stocks, increasing restrictions on where, when, and how fishing can be conducted had been placed almost in every country. The distinction between straddling stocks and highly migratory fish stocks is that the former stay mainly within the EEZ of one country and migrate out of the zone only occasionally. The highly migratory fish stocks instead migrate over long distances and can be found both in the EEZ of several countries and in the high seas. Apart from these two main agreements, a considerable number of other international legal instruments are operational, dealing more specifically with the preservation and conservation of living marine resources. The major ones are:

- the Convention on the Conservation of Antarctic Marine Living Resources (CCAMLR);
- the International Commission for the Conservation of Atlantic Tunas (ICCAT);
- the Commission for the Conservation of Southern Blue-fin Tuna (CCSBT);
- the Code of Conduct for Responsible Fishing (CCRF), issued by FAO in 1995. This is a voluntary—thus not binding—code which contains principles for sustainable fisheries governance, followed by four International Plans of Action (IPOAs on sharks, seabirds, IUU, and fishing capacity), a Strategy for improving status and trend of capture fisheries, and 47 technical guidelines;

- the Reykjavik Declaration issued in 2001, which introduces the ecosystem approach;
- the Compliance Agreement issued in 2003, addressing the problem of re-flagging of fleets by giving a black mark to vessels contravening rules on international conservation. A national record of the fishing vessels authorized to fish on the high seas is required from the parties to the Compliance Agreement.

Furthermore, there are national fishery administrations with legal authority over both fishing and the related activities at national level (where about 90 percent of global fishing takes place), and numerous (about 20) Regional Fishery Bodies (RFBs) located worldwide with a mandate for marine fisheries, plus over 30 regional fishery conventions.

Measures aiming to implement fish stock preservation include the reduction of the fishing fleets, temporary closing of areas, exclusive use rights, season or species restrictions, eco-labelling, minimum fish size and gear restrictions. However, the main problem is yet to monitor and enforce the present measures. In 2005 the FAO Committee on Fisheries, in order to assess the progress in the implementation of the CCRF and the related IPOAs, started to collect information on a self-assessment questionnaire basis (FAO 2005). Overall, out of 49 FAO member states (27 percent of the FAO membership) that responded to the questionnaire, one in five members reported not to have any fisheries management plans in place at all. Among the management tools applied by member states to implement their management plans, the most commonly reported was the prohibition of destructive fishing methods by addressing the selectivity of the fishing gears. The least used tools were control mechanisms such as the use of stock specific reference points, product traceability, and trade instruments. With regard to by-catch and discard data, just over half of FAO member states reported to monitor these parameters on a regular basis. Global fish and other marine organisms discarded by-catch has been estimated at over 20 mmt per year (FAO 2000).

Despite being badly governed, marine fisheries provide vast global benefits. In terms of jobs, primary sector employment in capture fisheries and aquaculture reached about 38 million people in 2002. Fisheries—along with processing and marketing on land—employs about 200 million people worldwide, 85 percent of which are concentrated in Asia (Botsford et al. 1997). However, even though fisheries are a direct source of work, food, and money for about 200 million people worldwide, from an economic perspective the capture fishery industry remains inefficient. A number of authors have examined the causes of fishery management failures. Some have focused on the lack of sufficient knowledge about biological and climatic mechanisms affecting fish stocks. Others have examined specific failures in decision-making mechanisms. However, few have attempted to examine the complex mixture of information affecting the capture fishery management decision-making processes. Fishery decision-making is a highly complex dynamic mechanism that includes elements of biological, economic, social, ecological, and physical aspects. An additional complication for decision makers is the inherent uncertainty that characterizes fisheries' data, the climatic variables affecting fish stocks, and the models used to estimate fish stocks.

Fisheries subsidies

As each country owns the resources within the boundaries of its EEZ, in principle the sovereign coastal state should gain an income from fisheries. The fish is a public asset and the one who takes it should pay a price to the state. Mining and offshore oil and gas companies pay the state to exploit these natural resources from the sea. Ironically, not only do states not charge fishermen for what they gain from public resources but the capture fishery industry also benefits by large government subsidies, making an already bad situation even worse.

Estimates of subsidies to fishermen vary greatly from USD 15 to 55 billion worldwide. FAO (1992) reported that subsidies in world fisheries exceed USD 50 billion per year, while a more conservative research study from the World Bank (Milazzo 1998) estimates the fisheries subsidies level to be between USD 15 and 20 billion per year. A report by the World Wildlife Fund (2001) estimates that subsidies in the fishery sector are at least USD 15 billion—about 20 percent of the value of global fish catches. Total subsidies for the OECD countries in 1997 were estimated to be USD 1.8 billion (OECD 1997). OECD classification of subsides includes MRE (Management, Research, Enforcement and enhancement), FI (fisheries infrastructure), IM (investment and modernization), TE (tax exemptions) and the 'Other' category (OT), which includes decommission subsidies, access to other countries' EEZ, income support and unemployment insurance.

The most common subsidies take the form of: grants; preferential credit lines and tax reduction for the construction of new vessels or the modernization of old ones; decommission payments (buyback of vessels); government expenditures to obtain access to other countries' EEZs; and reduced prices or tax for input purchasing and income/wage supporting policy. Historically, these payments were initially transferred to producers during cyclical downturns, however, once instituted they have become very difficult to remove.

Subsidies perversely affect economic decisions creating distortions that cause prices to neither reflect the resource scarcity nor the cost of production and consumption. A price support policy—increasing the price of fish—would raise total revenue proportionately. In open-access fishery the increased price of fish temporarily leads to higher profits in the industry.

This, in turn, leads to new entrants and, therefore, to a higher level of effort. Fishermen continue to enter the industry until profits are driven to zero again.

Subsidies in the form of grants or tax reduction lead to a downward shift in the fishing industry's long-run cost curve (see Figure 6.3). This, in turn, induces an expansion of fishing capacity and efforts until costs and revenues are once again equal, but at a higher equilibrium point resulting in over-fishing. Hence, since prices do not reflect the real value of the goods, adverse effects arise. From the entrepreneurial side, subsidies fuel investment leading to overcapitalization of the entire sector, while from the consumer side subsidies stimulate over-consumption. As a result, the harvesting capacity of the fleets is far larger compared to the size of the fish stocks and over-fishing is exacerbated, for it is achieved faster and maintained over time, even when stocks are depleting. A different situation would arise if subsidies were paid to contract fishing capacity rather than to expand it, easing the problem of overcapitalization and over-fishing.

Although the result of unintended consequences, capture fishery subsidies have not only distributional effects but also adverse impacts on fishery management, on the environment and on biodiversity.[35] Moreover, while subsidies may benefit certain groups in the short term, they ultimately cause distortions that lead to economic inefficiency and prevent environmental costs from being internalized. Despite their negative effects on the environment and on biodiversity conservation, governments continue to allocate subsidies to the capture fishery sector. Recently, the environmental implication of subsidies was being explored and debated internationally at the WTO Doha Round and a ministerial declaration has since been released. One of the major barriers to reform subsidies is the powerful lobbying of influential interest groups that are benefiting from these policies.

CONSERVATION PROBLEMS

Fishery management faces three main problems related to conservation: (1) illegal unreported and unregulated fishing (IUU); (2) harmful fishing; and (3) invasive alien species (IAS) damages. These issues are gaining increasing international attention both in the scientific and the policy arenas, since they affect fish stocks overexploitation and depleting, by-catch, and even provoke destruction of marine habitats, ecosystems, and biodiversity.

IUU fishing refers to fishing activities taking place outside the management and regulation framework for a particular fishery—e.g. fishing out of season or fishing in coastal state EEZ without permission. One of the consequences associated with IUU fishing is the erosion of sustainable management practices for the target species. Vessels flying a 'flag of convenience'—i.e. vessels registered in states that did not ratify fishing agreements—conduct most of IUU fishing. IUU fishing accounts in some countries as much as 30 percent of the total catch. An evolving norm in fisheries management is the use of landing and trade controls, such as documentation, labelling and certification requirements in order to avoid or at least reduce IUU catches.

Another conservation problem in fisheries is the use of equipment lacking the necessary environmental safeguards (harmful fishing). There are catching methods that are harmful to non-targeted species and to the seabed. As a consequence, the development of alternative fishing technologies aims to increase selectivity and to reduce environmental damage.

Harmful fishing practices include dredging, long-lining, fine-mesh nets, cyanide, and even the use of explosives. With such ecologically detrimental fishing gears the damage is far bigger than the revenue, but the environment costs are hardly ever internalized in open-access fisheries. Two widely used methods of harmful fishing are trawling and long-lining. The two methods sort different size and quantity of fish caught. Trawling is an indiscriminate way for fishing at relatively shallow depths and on fairy smooth grounds that destroys benthic communities and the seafloor. Long-liners tend to catch larger fish in deeper waters (1,000 to 1,500 m) exploiting fish that are unavailable to trawlers.

Another harmful fishing method is large-scale driftnets. Driftnets are made by multiple polyamide filaments and are composed of successive 100 m long elements linked together to reach variable lengths. Once lowered into the sea, they float on the surface by means of a headline and are kept vertically by another weighted headline 20 m below, so that the net hangs downward. The driftnet is a highly non-selective method of fishing—by-catch figures for this method are reported to be up to 15 percent (Eceizabarrena Saenz 2001). Both high-sea trawlers and driftnet vessels usually operate at night any time of the year with small crews whose black market wage is paid on a shared-out basis. They can set 15–50 km of driftnet per night. Trawling on the seafloor represents a serious threat to the biodiversity of the marine ecosystem. It has been estimated that the world's trawling grounds could approximately total 20 million km^2 (WSSD 2002). Although a number of legal instruments have been adopted in order to stop driftnet fishing, such as the UN General Assembly Resolution 46/215 adopted in 1991, this harmful practice continues.

Another problem linked to harmful fishing is by-catch. By-catch is usually composed of undersized and unmarketable fish, which is mainly the consequence of insufficiently selective fishing methods and gears. It affects non-target species, such as for instance the killing of dolphins in tuna fishing and the accidental death of turtles in the trawl nets used to catch shrimps. The Endangered Species Act limits commercial access to fishery resources according to the potential impact fishing activities may have on endangered or threatened species. FAO (1994) estimated that by-catch amounts to 27 mmt worldwide each year. Even if FAO is now recording a decrease in the total discards, incidental catch still remain high.

Recent technological devices are intended to keep some animals away from certain areas. These include Acoustic Harassment Devices (AHD), which use sounds to put marine mammals off following commercially harvested species, and Turtle Excluder Devices (TED), to prevent turtles from passing into the inner part of the seine. The TED is a trap door through which turtles can escape. In addition to the impacts of over-fishing of targeted species and incidental catch, harmful and large scale fishing techniques and equipments have a huge ecological impact on the marine environment in terms of mortality rate of non-commercial fish and destruction of marine habitats and bio-diversity. A 1980 US law requires shrimpers to fit their nets with TEDs. Another example to limit by-catch is the introduction of a bird-scaring device, which helps reduce by-catch of sea birds in long-line fishing.

IAS is an externality associated with globalization and international trade and it is likely to be one of the main causes of biodiversity loss worldwide. Bio-invasion occurs when an alien organism is introduced into a new environment. The invading organism may significantly modify ecosystem's balance and species interactions. The major pathway of marine bio-invasion is from ballast water discharge from ships and from hull fouling organisms. A consortium led by the University of Bergen in Norway is currently developing the system Ocean Saver, which is already operational using a combination of N_2 and filtering.[36] Other ways of treating ballast water include microwave treatment or radiation. In addition, there is an international IMO convention on ballast water—of course, this is only applicable to the countries that signed the convention—under which newly built ships should undergo a treatment facility starting from 2008 and all international seagoing ships should have a treatment plant built in the ship starting from 2016.[37] Presently, there are no practical tools to avoid IAS introductions by hull fouling organisms (POGO 2001). It has been estimated (WSSD 2002) that each year about 12 billion tons of ballast water containing—at any one time—3,000 marine species are shipped around the globe, spreading worldwide alien and invasive species.

7 Marine Bioproducts

MARINE-DERIVED PRODUCTS

Nature has been the traditional source for organic chemical compounds used in medicine for over 3,000 years. The oldest known pharmacological script mentioning marine cures is the Chinese *Pen Tsao*, which was written around 2800 B.C. It recommends seaweed-based recipes for a number of disorders such as pains, abscesses, menstrual difficulties and cancer. The Romans attributed medicinal effects to some marine animals to treat infections, inflammation, arthritis and other diseases. Seaweed remedies are still used by the San Blas Islands Indians in Panama (Jimeno et al. 2004).

Tyrian purple was the first marine natural product to be used for commercial applications. This renowned dye of antiquity dates back to the 13th century B.C., when the Phoenicians used to extract this product by the hypobranchial glands of a marine mollusc called *Murex brandaris*. Tyrian purple was one of the most expensive commodities in the ancient world, due to its low abundance in the marine animals from which it was extracted. To produce just 1.4 g of pure dye, the collection and extraction of approximately 12,000 molluscs was required. Tyrian purple was also the first marine metabolite whose structure was correctly deduced and proven in 1909 (Rohan 2000).

In the 17th century scientists began to isolate, purify and define natural substances used since ancient times. However, apart from sporadic early works at the beginning of the 20th century, little research was performed on the chemistry of marine creatures up until the 1960s. This can mainly be attributed to the difficulties associated with the collection and identification of the marine organisms. The oceans started to attract interest from pharmaceutical companies and research institutions only about 50 years ago, when the sponge-derived nucleosides *spongothymidine* and *spongouridine* were first discovered. Since that time, this relatively new scientific field has grown considerably, especially in regard to drug discovery and development. The chemistry of organisms such as *phytoplankton*, macroalgae, sponges, coelenterates, bryozoans, molluscs, tunicates (or ascidians) and echinoderms has been extensively studied.

From a scientific perspective, the past 40 years have seen thousands of marine metabolites isolated, many of which have been structurally novel and unprecedented within the terrestrial biosphere. Novel metabolites have been isolated from each of these taxonomically distinct groups; however, the distribution of novel compounds amongst each group has been unequal, with some groups possessing more interesting compounds than others.

Research efforts up until the mid-1980s predominantly focused on the algae, sponges and coelenterates and accordingly these organisms contributed to the majority of the literature published on new marine metabolites. The identification of significant biological activities of tunicate extracts for various pharmacological or agrochemical applications during the last fifteen years contributed to focus scientists' attention on tunicate chemistry.

Metabolites produced by marine bacteria and invertebrates have yielded pharmaceutical products such as novel anti-inflammatory agents (e.g. pseudopterosins, topsentins, scytonemin, manoalide) anti-cancer agents (e.g. bryostatins, discodermolide, eleutherobin and sarcodictyin) and antibiotics (e.g. marinone). Antioxidant peptides have been isolated from protease digests of prawn muscle and the mycosporine amino acid (MAA) precursor 4-deoxy-gadusol has been isolated from seaweeds and micro-algae. These peptides have applications as food additives and in cosmetics.

Many other marine-derived products other than drugs are already on the market (see Table 7.1). The past two decades have seen a blossoming of applications in marine-derived cosmetics, nutraceuticals and enzymes. Each of these classes of bioproducts has a potential multi-million dollar market value. For example, the Estèe Lauder anti-wrinkle cream product Resilience® contains a Caribbean gorgonian[1] extract called pseudopterosin E as the key ingredient responsible for reducing wrinkles. Infant-formula makers employ algae-derived substances in some of their products. For instance, two essential fatty acids present in human breast milk are also manufactured by the marine micro alga *Crypthecodinium cohnii*. The Martek Biosciences infant formula Formulaid is a nutritional supplement that contains both the fatty acids isolated from this marine micro algae and a DNA polymerase that was isolated from a deep-sea hydrothermal vent bacterium. The anti-irritant pseudopterosin, which active agent is extracted from the soft coral *Pseudopterogorgonia elisabethae* (a sea whip) is used in skin-care beauty products (Committee on Marine Biotechnology 2002).

An enzyme synthesized from hydrothermal vents microbes has proved highly effective in decreasing underground oil viscosity and therefore is used for increasing oil well yields. Natural glues produced by the common blue mussel are used in chemical compounds to improve the adherence of the automobiles' paintings, in sutureless wound closure and dental fixatives. New varieties of artificial bone grafts, produced from ground-up corals, possess a porosity that precisely mimics that of human bone tissue and can, therefore, have applications in orthopaedics.

According to a survey conducted by Newman et al. (2003), 61 percent of the 877 new chemical entities introduced as drugs worldwide between 1981 and 2002 were natural products, natural product derivatives or natural product mimics. Percentages were even higher when considering only the antibacterial (79 percent) and anticancer (74 percent) compounds. Three of the most common examples of derivatives from natural products include aspirin, morphine and penicillin. Although still very few marine-derived pharmaceuticals have become marketable drugs, there is great potential that many marine natural products presently under investigation will eventually do so. The increasing number of patents in recent years witnesses evidence of the growing industrial significance of marine natural products. The size of pharmaceutical and personal care applications respectively had a market value of USD 330 and 64 billion in 1999 (Committee on Marine Biotechnology 2002).

ORGANISMS BIOCHEMICAL PROPERTIES

In terms of evolution and biodiversity, the sea appears to be superior to the terrestrial ecosystem. Although the first living organisms appeared in the sea more than 3,500 MA ago, the most important biological explosion took place in the marine ecosystem during the Cambrian period (600 MA ago).

The environmental and ecological challenges facing marine organisms, such as fluctuations in salinity and pressure, and the effects of predation, bacterial and viral pathogens and the intra- (amongst members of the same species) and inter-specific (between different species) competition for limited resources—e.g. nutrients, food and space—has led to a very diverse set of evolutionary adaptations, developed over the past millions of years. One of these adaptations is the ability to synthesize and release a huge diversity of chemical agents. Among these, there are molecules with powerful biochemical properties developed in order to survive in this highly competitive environment.

A number of sessile organisms—i.e. organism being attached to a substrate and not free to move about—produce natural products as their defence mechanism. Lacking any physical protection from predators, being soft bodied and having a sedentary lifestyle, they must protect themselves through chemical means of defence. Therefore, they have evolved the ability to synthesize toxic compounds or to obtain them from other marine microorganisms. Moreover, many invertebrates such as sponges, corals and tunicates feed by filtering water and since seawater contains high concentrations of bacteria, filter-feeding invertebrates must produce antibiotics themselves in order to survive.

During the past 30 to 40 years, marine researchers have discovered a large number of unique and chemically diverse bioactive natural products

Table 7.1 Examples of Products Derived from Marine Organisms

Products	*Specific product*	*Source*	*Uses*	*Enterprise / trade name*
Algal	Carrageenans	Red algae	Cosmetics Thickener	Marine colloids (USA)
Polysaccharides	Agars Alginates		Pharmacy Mucoprotector Anti-coagulant Antiviral	Danisco (Denmark) SOBALG (France)
Glycosamino-glycans	Chondroitin sulphate	Fish	Cosmetics Tissue replacement Anticoagulant	CTTP (France)
Collagen			Cosmetics Artificial tissue	
Chitosan	B (1-4) N-acetyl Glucosamine	Crustacean Shells, Fungi	Cosmetics, colloids Pharmacy, Microencapsulation	
Lipids	Long chain PUFA (AA, EPA, DHA)	Microalgae Seaweed Fish	Prevention of heart disease, mental development in premature children Antitumoural Lipid metabolism	BIONAGROL 3000 AGE OMEGA 3 (Arkopharma) MAXEPA (Pierre Fabre Medicament)
Peptides	Hormones, cyclic Peptides	Fish Hydrolysates	Antioxidant Immunostimulants Nutraceutical products	Promarine

Source: European Science Foundation (2001) Marine Biotechnology: A European Strategy for Marine Biotechnology. Reprinted with permission.

coming from a broad variety of marine organisms, including microorganisms, invertebrates and algae. Some of these marine organisms contain unique natural toxic chemicals that might also attack human cancer cells. As a result, over the past 30 years scientists have extracted at least 20,000 new biochemical substances from marine creatures, dozens of which have already reached clinical trials (Krajick 2004). Between 1969 and 1995, approximately 200 new patents were issued worldwide for marine-derived biochemical products with potential therapeutic activity. Between 1996 and 1999 the rate of discovery and patenting increased even more, with around 100 new compounds patented in just 3 years. So far, the US National Cancer Institute (NCI) has screened tens of thousands of marine extracts and the institute continues to receive roughly 1,000 marine organisms each year. Unfortunately, only a few of the hundred compounds reaching the preclinical testing stage yield a commercial potential for pharmaceuticals. Although the marine bioproduct sector is not yet commercially developed, there are a number of reasons to believe that it will be in the near future.

1. Given the richer biological and chemical diversity of the oceans compared to land, the marine environment has an enormous wealth in terms of bioproducts. However, the major biodiversity in the oceans is in its microbial life. One millilitre of seawater contains around 1 million microorganisms.

   Even though marine microorganisms can be a source of novel bioproducts, 99 percent of the microorganisms in the ocean have not yet been studied for potential commercial applications (Committee on Marine Biotechnology 2002). The successful trend of terrestrial microorganisms in the development of new drugs is a strong incentive for a systematic investigation of marine microbes. The research priority in this field is to discover and identify bioactive molecules from marine biota that can have useful applications. Marine genetic resources can have applications in pharmaceutical, agrichemical and commercial sectors. Examples are the sustainable development of the aquaculture sector, the creation of bioremediation agents for environmental protection and bio-antifoulants, and numerous other industrial applications.
2. Advances in diving technology are opening new deep underwater territories to bioprospecting, i.e. the search and discovery of exploitable biology. Marine bioprospecting has focused on tropical waters because of the higher biodiversity concentration in this area. However, marine habitats at other latitudes must be explored as well since they may contain new species to discover.
3. Recent developments in molecular biology and genetics allow labs to isolate molecules much more rapidly than before. As a result, the oceans' potential as a biochemical resources reservoir is becoming more and more important. A project led by Craig Venter, one of the

scientists who mapped the human genome, intends to sample the DNA of bacteria from the oceans. Given the advances in bioinformatics, the question is whether it is really necessary to positively identify the organisms from which to isolate a novel bio-compound. In the future it will be possible to map an enormous amount of different genes from marine organisms, group them in functional groups and clone them into bacteria for production.

There are important biological communities of marine organisms—called chemiosynthetic[2] biota—found at specific marine habitats such as cold seeps, trenches and hydrothermal vent sites. The hydrothermal vent genesis and main physical features have already been described in a previous chapter.

Deep-sea hydrothermal vents and their associated fauna were serendipitously discovered in 1977 during a survey along the Galapagos Spreading Centre (Ecuador) in the Eastern Pacific aboard the research submersible *Alvin* at an approximate depth of 2,500 m (Van Dover et al. 2002). The fauna associated with these extreme environments is relevant among marine resources as potential genetic material for human drugs and other applications. Numerous scientific expeditions indicate that significant lengths of Ridge crests are hydrothermally active and that complex biological communities are common to these areas. In the framework of the project Census of Marine Life, the main idea behind the under-project ChEss (Biogeography of Chemosynthetic Ecosystems) is to explore the fauna and flora of hydrothermal vents located in the North Atlantic using an AUV equipped with chemical sensors. The project, which testifies to the interest of the international research community in this field, is still under way and will undoubtedly shed light on these unique organisms.

Deep-sea biological communities living near hydrothermal vent areas are commonly called *thermophiles*—heat lovers—and are composed of bacteria (*hyperthermophilic* microbes), massive giant tubeworms (*vestimentiferan, Riftia pachyptila*), blind crabs (*Bythograea thermydron*), huge white clams (*vesicomyd*), giant mussels (*bathymodiolid*) and white eyeless shrimps (*Rimicaris exoculata*).

Thermophiles have developed physiological and biochemical adaptations to extreme environments as hot as more than 400° C and subject to the immense pressure of the depth. All larger hydrothermal organisms are dependent on the *extremophile* bacteria, which are able to synthesize organic compounds such as carbohydrates by oxidization of the hydrothermal vents' toxic hydrogen sulphide (chemosynthesis). Thus, the primary producers of the food chain of the vents' ecosystem are bacteria. These micro-organisms (cells without a nucleus) sustain the local macro-fauna (invertebrates) either living inside them as endosymbionts, as in the case of the abundant tube worms, or constituting the primary food source by making large microbial mats surrounding the vents, which, in turn, are eaten by gastropods, shrimps etc. The knowledge of these organisms' existence, which is a unique example of chemosynthetic life on this planet, has revolutionized the understanding of where and how life may occur and has stimulated new theories of the origin of life. This is one of the few complex ecosystems known that live on energy from chemicals, converting hydrogen, hydrogen sulphide and methane coming from the high-temperature fluids into vital energy. More than 500 new species have been described from vents areas in the past 25 years. Exploration of the Ridge system in different ocean basins has confirmed regional variations in species composition, even though the ecological, phylogenetic[3] and geochemical factors that contribute to the distinct evolutions of the vents' fauna are still under exploration.

Similar chemosynthetic ecosystems were subsequently found at cold seeps, where more than 200 morphological species had been described since their discovery. Cold hydrocarbon seeps, sometimes called 'cold vents', were discovered in 1983 at the base of the Florida Escarpment. Occurring along passive and active continental margins, cold seeps result from seepage of cold fluids with high concentrations of methane or sulphide from the underlying sediments and support a chemosynthetic life similar to that of the hydrothermal vents. Although cold seep areas are similar to hydrothermal vents, they differ because the emitted fluids are about the same temperature as the surrounding waters. This is the reason why they are called 'cold' seeps in contrast to the extremely hot fluids coming from hydrothermal vents.

The biological communities surrounding cold seeps are—as is the case for hot hydrothermal vents—dependent on bacterial chemosynthesis, made by groups of micro-organisms belonging to both the *eubacterial* and *archea* domains of life. However, the chemicals found in the cold-seep fluid—such as hydrogen sulphide, methane, iron, manganese and silica—are different from those of hydrothermal vents' and, therefore, the characteristics of cold seeps biological communities are different as well. Although some vent and seep invertebrates are relatively young, the origin of some of them may be very ancient, extending back to Palaeozoic (between 540 and 248 MA ago) and Mesozoic (245 to 65 MA) (Van Dover et al. 2002).

Other marine areas where unique genetic material can be found are trenches, where sunken wood, whale carcasses and kelp fall and accumulate in the seabed. Decomposing whale skeletons release sulphide through microbial reduction of sulphates, sustaining sulphide-oxidising bacteria and communities of invertebrates based on chemosynthetic production (ChEss 2003). Data collected during numerous dives of submersible *Alvin* in the Santa Catalina Basin between 1987 and 1988, indicate that the whales' remains on the seabed produce a habitat (mussels, snails and limpets) distinct from the surrounding ocean floor but similar to the hydrothermal vents' and cold seeps' ones (Smith et al. 1989). Similar biota, such as *osedax* worms, have been found on whales' bones at other sites, suggesting that microbial degradation of organic-rich whale remains yields sulphide from sulphate reduction and putrefaction, thus supporting animals that depend on chemiosynthesis.

These organisms provide new materials for use in biochemical research. There is a whole spectrum of chemiosynthetic organisms that can have great potential for pharmaceutical, industrial and environmental applications. Almost 6,000 bioactive metabolites from hydrothermal vents microorganism have been described in recent literature. Drugs extracted from deep-sea bacteria are exceptionally valuable because of the extreme power of their toxins, useful in the fight against cancer, infections and Alzheimer disease. Some of these organisms also contain valuable enzymes that have commercial applications in chemical compounds employed in hazardous waste management and food processes.

To collect vent deep-sea microorganisms from the seafloor, the scientists use underwater vehicles. Organisms are then processed in laboratories where their isolation from the mud and cultivation can be performed under pressure and temperature conditions similar to the deep-sea environment. Eventually, the biomaterial is sent to the commercial sector.

Recently, attention has been focused on *hyperthermophilic*[4] bacteria which, living in temperatures as high as 110° C, possess thermostable properties that can be used in industrial processes involving high heat and pressure. Other forms of microbes have been found in cores recovered from several hundred meters below the sea bottom. This discovery led scientists to believe in the existence of a biosphere living in the deep sub-seafloor. The recent discovery of the abundance and distribution of deep, cold-water corals is another example of the ocean potential for biogenetic material (National Research Council 1995).

An estimate of the market potential of these bacteria, reports a budget of USD 3 billion per year (Glowka 1996). There is concern about the negative environmental impact of mineral mining in hydrothermal vent ecosystems, as a conflict between the exploitation of hydrothermal bio-resources and deep seabed mining of sulphides may occur. As a matter of fact, the limited knowledge about *extremophiles* does not permit a sound assessment of the impact that seabed mining can provoke on hydrothermal fauna. However, precaution is recommended, even if there is evidence that vent communities are ephemeral because a hydrothermal system is inherently a short-term phenomenon. Eventually vents plug themselves up and the hot water that sustains these communities will stop.

MARINE BIOTECHNOLOGY

The word biotechnology was coined in 1919 by Karl Ereky to apply to the interaction of biology with human technology. Biotechnology is usually defined as the means or the way of manipulating life forms on a molecular level, but also as the utilization of knowledge pertaining to living systems to provide desirable products for man's use. The OECD (1989) definition of biotechnology is "The application of science and technology to living organisms, as well as parts, products and models thereof, to alter living or non-living materials for the production of knowledge, goods and services".

The development of molecular genetics started in the 1970s and it presently involves both the manipulation of genetic material (DNA) and the transfer of genetic information between different species. Recombinant DNA is an artificial DNA sequence resulting from the combination of different DNA sequences. Gene splicing is one of the most common molecular genetic techniques, pieces of DNA from one or more organisms are transferred via a 'vector' into a different organism that integrates them into its genome—which is the entire set of genes of an organism. In this way the receiving organism obtains new capabilities for biochemical transformations. Another widely applied technique is cell fusion—or hybridoma technology—in which two cells of different types are fused, producing a strain containing genetic materials of both cells (Cicin-Sain et al. 1996).

According to OECD (2004) biotechnology research adopts five different technologies or processes:

1. DNA coding, which includes genomics, pharmaco-genetics, gene probes, DNA sequencing/synthesis/amplification and genetic engineering;
2. Protein and molecule manipulation, which includes protein/peptide sequencing/synthesis, lipid/protein engineering, proteomics,[5] hormones and growth factors, cell receptors/signalling/pheromones;
3. Cell and tissue culture and engineering, including cell/tissue culture, tissue engineering, hybridisation, cellular fusion, vaccine/immune stimulants, embryo manipulation;
4. Process biotechnology, which includes processes such as bio-reaction, fermentation, bio-processing, bioleaching, bio-pulping, bio-bleaching, bio-desulphurisation, bioremediation and biofiltration; and
5. Sub-cellular organisms, including gene therapy and viral vectors.

Although in the past biotechnology relied primarily upon the exploitation of terrestrial organisms, vent biota and other novel marine organisms recently discovered have brought in the arena the development of marine biotechnology. Hydrothermal bacteria and marine genetic materials are already used in biochemical synthesis and bioengineering applications, and some pharmaceuticals and other industrial products are already marketed. The economic value of these organisms is expected to be of huge significance in the future, since scientific knowledge in this field is expanding at an exceptional rate.

The understanding of life during the last years, thanks to the advances of genomics, molecular biology and bioinformatics at cellular and genetic level, has opened new perspectives in studying and exploiting marine organisms. This has resulted in the establishment of research centres focused on marine biotechnology. The main objective of scientists is to develop novel drugs for treating human diseases and producing diagnostic devices for monitoring health. However, marine biotechnology has also several other applications, such as techniques to monitor and assess, restore, protect and manage marine environments, ensuring sustainable and safe aquaculture and fisheries, and discovering new types of composite materials, biopolymers and enzymes for industry (ESF 2001). Moreover, marine microorganisms frequently produce environmentally friendly chemicals such as biopolymers and non-toxic biosurfactants, which can be applied in environmental waste management and treatment.

Four major applications of marine biotechnology have been identified (Cicin-Sain et al. 1996):

1. Aquaculture and seafood enhancement. Marine biotechnology has been applied to aquaculture to improve fish resistance to illness, temperature tolerance and disease control, to increase productivity by obtaining fast-growing species of bigger size (transgenic organisms) and also to seafood enhancement;
2. Commercial and industrial application. Biotechnology based on marine organisms has been applied in trying to obtain products such as biological anti-biofouling agents and adhesives for the shipping and maritime industries.[6] Some bacteria are real living catalysts, which can be used, just to name few industrial processes of potential application, in high-temperature industrial applications and in the field of treatment of hazardous waste. For instance, polysaccharides derived from marine crustaceans are used as principal agents to control ice formation in frozen foods and as antifungal agents for agriculture;
3. Biomedical and pharmaceutical products. The main applications in this field concern pharmaceutical products such as genetically engineered medicines and vaccines, cosmetics and nutraceuticals. The ongoing applications are so many that an entire section is dedicated to them. Presently, the most widely commercially used enzyme to replicate DNA is derived from an enzyme found in microbes living in hydrothermal vents areas. In fact, two *thermophiles* species—*Thermus aquaticus*[7] and *Thermococcus litoralis*—are used as source of the enzyme DNA polymerase for the polymerase chain reaction in DNA fingerprinting (Mgbeoji 2004). The ancient horseshoe crab (*Limulus polyphemus*) supplies blood used in common toxin-screening tests, while the nerve cells of the long-finned squid (*Loligo pealei*) include 'giant axons', which are used by neurobiologist scientists as an analogue to understand mammalian neurobiology. *Discodermolide*, a compound extracted from marine sponges, has been shown to stop the growth of cancer cells in laboratory tests (Committee on Exploration of the Seas 2003);

4. Environmental monitoring and resources management. Environmental biotechnologies attempt to minimize pollution and enhance materials and energy use, while maximizing production of recyclable or biodegradable products. Enzymes from marine microorganisms that can function at extreme pH or temperature are being isolated and bioengineered to degrade hazardous wastes. *Bacillus stearothermophilus*, another vent biota that can sustain high temperature, is used to obtain enzymes that find application in biological washing powders.

An estimate of worldwide sales of marine biotechnology products in 2000 reports a value of USD 100 billion, while the potential market value of the industrial use of these enzymes has been assessed at USD 3 billion per year (Gorina-Ysern 2003).

MARINE BIOMIMETICS

Animal natural technology is sometimes way ahead of human achievements. Biomimetics is the new science studying animal technology and its applicability to human technology; an example of the fact that the present technology cannot yet compete with the sophisticated systems possessed by some biota comes from sponges. Sponges have yielded promising compounds, such as the chemical extracted from *Tethya crypta*, which is used in antiviral medicines that presently have over USD 50 million in annual sales (Sanchirico 2004). Since 1996, Italian scientists discovered that deep-sea sponges in Antarctica's Ross Sea use tiny, needle-like glassy spicules embedded in their tissues to convey sunlight deep into their inner chambers (Cattaneo-Vietti et al.).

The deep-sea "glass" sponge *euplectella* provides another good example of advanced animal technology. It has fibre-optical properties surprisingly similar to those of commercial telecommunication fibre; however, since this sponge has some technological advantages over the man-made version, it is shedding light on how to make better fibre-optic cables (Sundar et al. 2003). The spicules of this sponge, which are even made of similar material of commercial telecommunication fibre, have a pure silica core with a higher refractive index than that of vitreous silica. Also the dimensions are comparable with that of fibre-optic, as well as the refractive index and, consequently, their wave-guiding transmission properties.

Other superior features of the sponge over the commercial telecommunication fibre are an effective crack-arresting mechanism and enhanced fracture toughness. The main weakness of commercial optical fibre is that they fracture because of their dependence on micro-crack growth. Another example of superior animal technology is the skin of the shark, which is made up of microscopic parallel grooves called "riblets" aligned in the motion direction to reduce drag and noise. The British Institute for Maritime Technology is trying to find a way to apply this animal technology to airplane and submarines on a commercial scale (Mann Borgese 1998).

Another desirable characteristic of many marine organisms is that they remain reasonably free of biofouling. Scientists have studied the characteristics of fouling-resistant animals and found out that they reduce fouling in different ways. Some do so by means of sacrificial slime, some by exudating films or tethered macromolecules on their surfaces, while others achieve these effects through the intrinsic low-surface-energy of their cellular and tissue surfaces (Committee on Marine Biotechnology 2002). The development of biofouling coatings based on these mechanisms has produced a significant reduction of drag.

Another example of sophisticated animal technology is the Hawaiian bobtail squid *Euprymna scolopes*, whose reflective tissues possess unique proteins—named "reflectins"—that provide a model of nanofabrication in animal system. Protein-based nanofabrication is a frontier area in biomimetics, which can inspire synthesis of nanostructured supramolecular devices, especially those used in spectroscopic and other optic applications. The structure of the protein found in this squid is engineered for possible applications in biomaterials, molecular devices and biosensors (Crookes et al. 2004).

The sea is also home to numerous macro-organisms donned of microporous materials, such as for instance the colonial coral *porites* or the sea urchin. Due to their unique properties (pore's size is uniform and ranges within a few millimetres), coral structures and urchin spines are used in vascular graft construction and orthopaedic surgery.

Highly interconnected microporous materials are still very difficult to produce synthetically. Employing these marine life-forms it is possible to fabricate materials that possess the same characteristics. The microporous material is replicated by using the calcium carbonate skeleton of the organism as a template. Microporous biomaterial has been primarily employed as medical devices and hard-tissue prostheses. The tiny size of the microporosity makes it ideal for making artificial organs and implants that become ingrown with host tissues. It can, therefore, be used to replace bone, blood vessels, trachea, and other damaged organs and tissues. Experiments conducted with dogs showed that after microporous ceramics were implanted, bone tissue regeneration grew into the pores of the material and mineralized (Committee on Marine Biotechnology 2002).

Another example of biomimetics comes from marine animals such as barnacles and mussels. These organisms persistently attach themselves to any kind of solid surface, natural or manufactured, wet or dry. The adhesives they use are extremely effective. Scientists have studied the active ingredient of these natural adhesives to develop new products for applications in science, medicine and industry. Being a protein, this adhesive gradually brakes down by natural processes within the human body and therefore it can also be used as an effective surgical adhesive.

PHARMACEUTICAL MARINE RESEARCH

During the past 20 years, thousands of novel compounds and their metabolites with diverse biological activities—ranging from antiviral to anticancer—have been isolated from numerous marine organisms. According to Arif et al. (2004) the marine pharmacy currently holds more than 35,000 marine-derived biological samples. About 150 compounds have cytotoxical properties against the tumour cells and approximately 35 compounds have a known mechanism of action for their antitumoural effect. Out of 35 antitumoural compounds, at least a dozen are currently under process in different phases of human clinical trials for treatment of different forms of cancer. The drug-discovery program includes extraction, purification and screening of natural materials to search for biologically active compounds. With the recent development of High-Throughput-Screening (HTS) analysis techniques a massive amount of material can be screened to identify a specific biochemical property in a much shorter time. Once the natural material has undergone the biological screening, its chemical structure must be elucidated. Analytical chemistry techniques include High-Pressure Liquid Chromatography (HPLC), Mass Spectrometry (MS) and Nuclear Magnetic Resonance (NMR) (Committee on Marine Biotechnology 2002).

Among other factors, in order for a marine organism to be considered for research, it must have a sufficient population, be suitable for clinical use and be economically obtainable. The primary area of concern about this new source of pharmaceuticals is the question of their supply. The organism must have a sufficient population that can be used for research and possible marketing because if the organism will be depleted shortly after a drug is derived, then it will be counterproductive to even begin the research cycle. A new anti-cancer drug typically takes 10 to 20 years to develop to the point of commercialization and has an average cost of USD 40 million. The chance that a new extract will yield a marketable drug is only 1 out of 250,000; therefore, the appropriate supply methods must be determined carefully prior to considering a marine chemical entity suitable for clinical development.

Many of the suitable bioproducts are scarce because they have a limited distribution or occur at a low biomass. Moreover, individual organisms often contain only trace amounts of the useful compounds. Thus, simply harvesting the source organism and extracting the active chemical is not a viable option because it may lead to depletion and extinction of natural populations. The ratio of compound vs biomass can be very low. For example, to obtain approximately 1 g of the promising anti-cancer agent ET-743, close to 1 mt of wet weight of *Ecteinascidia turbinata* have to be collected and extracted. Regarding Halichondrin B, a powerful cytostatic polyketide of sponge origin, only about 300 mg of agent can be obtained from 1 mt of the sponge *lissodendoryx*.

When it is not possible to harvest a sufficient quantity of the organism directly from its natural environment, supply must be ensured in a different way. Effective production methods for these organisms must therefore be developed. To fulfil this need, research is involved in developing a technology to supply the pharmaceutical market with sufficient amounts of marine chemicals. Production of sufficient amounts of the biologically active compounds depends on three factors:

1. The capacity to culture the organism in large quantities;
2. The ability to chemically synthesize the compound;
3. Genetically manipulate culture systems to produce the compound by means of modern molecular biological techniques (cloning of genes).

Culture methods provide one way of controllable and clean production, coupled with little environmental impact—since harvesting from natural habitats would be minimal. Mariculture, which has already been successfully developed for fish and shrimps, has achieved considerable progress with the bryozoan *Bugula neritina*—the source of the bryostatins—and with the tunicate *Ecteinascidia turbinate*—the source of ET-743.

Since many marine invertebrates are filter feeders that ingest bacteria or other small particles from the inhaled seawater, some bioactive metabolites previously isolated from sponges or tunicates are in fact of microbial origin and are taken up through the food chain. If the pharmacologically active natural product obtained from sponges or other invertebrates can be traced to exogenous sources such as bacteria or cyanobacteria and the gene is isolated, a sufficient production of the desired metabolites can be supplied by fermentation. Studying these microorganisms, during a three-year project conducted at Oregon State University with a panel of 2,500 cultures of marine bacteria, it has been assessed that less than 10 percent of the microbial life can be grown using common method of microbial cell cultivation (agar plating techniques) (Committee on Marine Biotechnology 2002). This limitation can be overcome with a further and even more advanced methodological approach for biotechnological production of pharmacologically important metabolites: cloning the enzymes responsible for the production of the drug into a bacterium host (e.g. recombinant e. coli).

Genetic manipulation to produce novel compounds that have potential utility for mankind is the last frontier of scientific research in this field and has incredible potential for creation of novel chemical entities. Cutting-edge biotechnology is molecular biology and genomics. While molecular biology investigates single genes, genomics is the sequencing, collection and interpreting of information contained within the genome (all the genes) of an organism. Microorganisms have been in the past the main target of genomic research because they have smaller genomes and are therefore more manageable. However, automated DNA sequencing (which is the use of fluorescence tracers instead of radioactive isotopes) and computational biology (or bioinformatics) have made it possible to rapidly sequence and annotate even larger and more complex genomes.

The capability to fully determine the genomic structure of an organism allows scientists to determine more quickly potential bioactive molecules, since the total DNA from a sample can be purified without first culturing the organisms. Cost for DNA sequencing has considerably decreased in the last decades, thus giving the possibility of using this approach to study new marine species and microbial populations that cannot be cultured. Complete genome sequences of more than 50 microbes have been published already. Sequencing of at least a dozen more is known to be complete but not yet published, while sequencing projects of about another 140 additional microbes are in varying stages of progress.[8] The benefit of inserting the genetic material into bacteria is that they can be fermented in large reactors and thus produce much more drug than would be available in nature.

Another important factor to consider in developing marine-derived drugs is the clinical viability of the active agent. One scale used to measure the clinical viability of the organism is to quantify its cytotoxic activity. Tetratoxins, the main derivative from most marine and some terrestrial poisons, are deadly to humans even in very small dosages. However, when these toxins are reduced to their active components, they can be used to produce very effective pain killer drugs. An NCI research conducted on 6,000 marine organisms showed that only 1.8 percent of them have a significant cytotoxic activity. The most promising marine organisms found thus far belong to three phyla: Porifera, Bryozoa, and Chordata.

Porifera are marine sponges. Sponges have been used medically for hundreds of years and many compounds from various sponge organisms have been isolated. More than 800 compounds have been identified in tropical sponges. One promising source of cancer-fighting substances is the compound called "bengamides", while another compound from a Caribbean sponge called *discodermia* is in clinical trials for the treatment of pancreatic and other forms of cancer.

One major finding in sponges resulted in a drug to cure both acute myeloid leukaemia and lymphoma. *Cryptotethya crypta*, a sponge living in the Caribbean, produces cytarabine, which is a compound used in a drug known as 1-beta-D-arabinofuranosylcytosine (ARA-C). Pharmacia & Upjohn Company markets this drug under the brand name Cytosar-UR, which is one of the very few marine anticancer drugs studied in long-term rat's cancer model for its efficacy (see Table 7.1). Other compounds coming from sponges are: Acyclovir—which treats herpes infections—and AZT—which fights HIV, the AIDS virus. These drugs can be traced to nucleosidic compounds that chemist Werner Bergmann isolated from *Cryptotheca*

Table 7.2 Examples of Commercialized Medicines from Marine Organisms

Chemical name	*Origin*	*Activity*	*Type of molecule*	*Year of commercialization*
Cephalosporins	Cephalosporium sp. marine fungi	Antibiotic	β-lactam	1965
Cytarabine (Ara-C)	Cryptotethya crypta Sponge	Antitumoural (cytostatic)	Nucleoside	1972
Kainic acid	Digenea simplex Red alga	Anthelmintic Insecticide	Amino acid	Early 1900s
Spongoadenosine (Ara-A)	Cryptotethya crypta Sponge	Antiviral Herpes	Nucleoside	NA
Ziconotide	Conus magus Mollusc	Analgesic	Peptide	1999

Source: European Science Foundation (2001.) Marine Biotechnology: A European Strategy for Marine Biotechnology. Reprinted with permission.

crypta in the 1950s. The Canadian company Inflazyme has developed another compound called contignasterol, which is the source of a potential treatment for asthma. This drug, based on a substance found in the Pacific sponge *Petrosia contignata*, reportedly produces fewer side effects than other current medications and can be swallowed instead of inhaled.

Bryozoans—a.k.a. "moss animals"—are aquatic organisms of the phylum *ectoprocta* living in colonies of interconnected individuals. Bryozoans are sessile animals with a lifestyle very similar to that of corals but, thanks to their unique body plan, they constitute a phylum of their own. Bryozoans' colonies range from millimetres to metres in size, however, single individuals are rarely larger than a millimetre. Some bryozoans encrust rocky surfaces, shells, or algae. Over 125 species of bryozoans are known to grow on the bottoms of ships, causing drag and reducing the efficiency and manoeuvrability of the fouled ships. Bryozoans may also foul pilings, piers and docks.

The marine bryozoans *Bugula neritina*, which lives off the California coast in the Pacific Ocean, is used for the production of an anti-cancer drug called Bryostatin-1. This compound is one of the most abundant and highly studied. Licensed by Bristol-Mayers Squibb, it is currently in Phase II of human clinical trial. Bryostatin-1 is a macrocyclic lactone that activates the protein kinase C, having antagonistic effects on tumour promoting. CalBioMarine, a Californian based company active in marine natural products genetic-based technologies, recently developed two novel compounds, of which one comes from a microbial symbiont of the Pacific coast arborescent bryozoan *bugula*.

Another potential field of exploration are tunicates. Tunicates are filter feeders that live both as sessile benthic animals and as passive floaters in the planktonic layer. They are placed in the phylum chordata. Tunicates belong to the sub-phylum *tunicata* or *urochordata*. *Tunicata*, which is a Latin word that means test or tunic, has two classes (sea-salps and sea squirts) from which an important compound is excreted.

Ascidacea, commonly called "ascidians", contains the majority of tunicate species. Ascidians can be either solitary or colonial in nature, with the largest solitary animals growing up to 15 cm. They are found mostly on hard surfaces such as rock, jetty pilings and coral, but they also grow on sea grasses and other vegetations. From a taxonomic perspective, tunicates are the most highly evolved marine animals routinely studied by chemists. Many compounds derived from tunicates are being tested in clinical trials. One of them is trabectedin, a compound that has been isolated from *Ecteinascidia turbinata,* a Mediterranean and Caribbean tunicate whose colonies look like translucent orange grapes.

To date a group of six ascidians-derived metabolites have reached antitumour clinical trials. The most notable examples of bioactive ascidians compounds include didemnin B, dehydrodidemnin B and ecteinascidin 743. The depsipeptide didemnin, which was isolated in 1981 from the Caribbean tunicate *Trididemnun solidum*, was the first marine-derived compound to enter clinical trials. In 1994 Phase II clinical trials sponsored by the NCI resulted in didemnin B, finally being dropped from further study due to its cardiotoxicity. However, aplidine (dehydrodidemnin B)—a second-generation didemnin that was isolated in 1988 from the Mediterranean tunicate *Aplidium albicans*—which is a closely related compound, has been found to be more active than didemnin B and lacks the cardiotoxicity of didemnin B. This agent demonstrated antitumoural activity against a variety of tumour models since it inhibits the synthesis of RNA, DNA and proteins. These results introduced Ecteinascidin 743 (Yondelis) into Phase I clinical trials in 1999. PharmaMar's (Zeltia group) in partnership with Johnson & Johnson Pharmaceutical Research & Development developed this product, which was first isolated from the Caribbean mangrove tunicate E*cteinascida turbinata* in 1990. Phase I clinical trials were successfully completed and ET-743 demonstrated to be particularly active against cancer cell lines derived from melanoma and carcinomas of the colon, breast, lung, brain and ovary (Amador et al. 2003). Phase II trials commenced in 1999 (see Table 7.3).

Another marine organism currently under investigation is a polychaete (marine tube worm) commonly called the "Pompeii worm". Built from the mineral secretions they produce, these sessile creatures form a vertical tube that provides a home within a few meters of a hydrothermal vent. A new anaerobic, heterotrophic and mesophilic bacterium classified as *Vibrio diabolicus* lives on the Pompeii worm in a symbiotic relationship. Studies conducted on bacteria have led scientists to believe that it may be used for regenerating bones damaged by osteoporosis and other degenerating diseases. The neuro-bioscience company International WEX Technologies Inc. has focused its resources on pharmaceutical uses of tetrodotoxin and

Table 7.3 Marine Natural Products and Derivatives in Clinical Development

Compound name — Source cemical class	*Company*	*Disease area*	*Status*
Compounds targeting-ion channels			
Ziconotide Cone snail Peptide	Neurex	Chronic pain	Phase III
AM336 Cone snail Peptide	AMRAD	Chronic pain	Phase I/II
GTS21 NemertineAnabaseine-derivative worm	Taiho	Alzheimer's disease Schizophrenia	Phase I/II
Compounds targeting enzymes			
Methionine aminopeptidase inhibitors			
LAF389 Sponge Amino acid derivative	Novartis	Cancer	Phase I
Protein kinase inhibitors			
Bryostatin-1 Bryozoan Polyketide	GPC Biotech	Cancer	Phase II
PLA2inhibitors			
OAS1000 Soft coral Diterpene-pentoseglycoside	OsteoArthritis Sciences	Wound healing, Inflammation	Phase I/II
Microtubule-interfering agents			
Dolastatin-10 Sea slug Peptide	NCI/Knoll	Cancer	Phase II
ILX651 Sea slug Peptide	Ilex Oncology	Cancer	Phase I
Cemadotin Sea slug Peptide	Knoll	Cancer	Phase II
Discodermolide Sponge Polyketide	Novartis	Cancer	Phase I
HTI286 Sponge Tripeptide	Wyeth	Cancer	Phase I
DNA-interactive agents			
YondelisTM Sea squir Isoquinolone	PharmaMar/ Johnson and Johnson	Cancer	Phase II/III
Oxidative stress inducers			
AplidinTM Sea squirt Cyclic depsipeptide	PharmaMar	Cancer	Phase II
Lysosomotropic compounds			
Kahalalide F Sea slug/alga Cyclic depsipeptide	PharmaMar	Cancer	Phase I
Immunostimulatory agents			
KRN7000 Sponge B-galactosylceramide	Kirin	Cancer	Phase I

Calcium-binding protein antagonists			
Squalamine lactate Shark Aminosteroid	Genaera	Cancer	Phase II
Compounds with unknown mechanism of action			
IPL512602 Sponge Steroid	Inflazyme/Aventis	Inflammation, Asthma	Phase II

Source: Haefner (2003). Reprinted with permission from Elsevier.

its derivative maculotoxin, coming from the puffer fish (a.k.a. blowfish). They are in the process of developing three products designed to aid in pain management: tectin; tetrodin; and tocudin. Tectin is in Phase IIa of its development.

Some researchers, including a group at Arizona State University, are investigating compounds called dolastatins for their potential anticancer properties. Dolastatins are peptides extracted from the sea hare *Dolabella auricularia*, which is a sea slug (a mollusc) living in the Indian Ocean. They believe that this organism has powerful chemical defences that might prove useful to people. Consequently, they began searching for the compounds using tried-and-true methods of biochemistry, such as grinding up specimens, dissolving the materials in various solvents and then testing the resulting extracts for a range of properties. Within this family, the linear peptide dolastatin 10 and the desipeptide dolastatin 15 exhibit the most promising antiproliferative actions and, as a result, were chosen for development. It appears that among their properties is the ability to kill bacteria, react with nerve cells or attack malignant cells.

Other products under development are based on agents isolated from different marine organisms, such as the sea slug *Elysia rufescens* and the clam *Mactromeris polynyma*. Another marine-derived drug that has been extensively tested for the treatment of chronic pain is Prialt, licensed by Elan Pharmaceuticals Inc. Prialt is a synthetic equivalent of a naturally occurring conopeptide (a venom) coming from a species of Pacific marine cone snail known as *Conus magus*. Prialt acts on nerve pathways to block pain more effectively than traditional opiates and appears to be 1,000 times more potent than morphine like stingers. To date, about 2,000 people have received the drug on an experimental basis and the drug has finally been marketed. Recently, a suite of uniquely active chemical extracts from a number of strains of cultured *thermophilic cyanobacteria* have been discovered. These extracts have shown a promising pattern of activity against resistant forms of cancer.

In spite of all efforts in marine natural products chemistry over the last three decades, except for the anti-viral drug Acyclovir,[10] AZT, the anticancer Ara-C, Priolt and the antibiotic Cephalosporin no marine-derived compound has been so far successfully launched into the market. Therefore, although marine biotechnology has demonstrated its applicability in producing pharmacologically active compounds, it has not yet matured commercially.

Clinical and commercial development of many marine natural products is idle for a number of reasons. Despite the success of numerous products of marine origin, the transition from the discovery to commercial development is not yet an efficient process. The discovery and commercial development of marine bioproducts is a complex, time-consuming, expensive and risky task. The process to market a novel product usually lasts five to seven years and can cost from USD 15 million to more than 500 million (Committee on Marine Biotechnology 2002). The development and marketing of bioproducts for human consumption are strictly regulated because of safety implications. Only a tiny minority of research-stage therapeutic compounds from natural sources enters clinical testing and only a small proportion of those reach the marketplace.

It has been predicted that isolation and structure elucidation will still play a major role in the future of marine natural products chemistry. Further advances in instruments will allow even smaller quantities of material to be purified and structures subsequently determined. The biological and chemical investigations of marine microorganisms have also been identified as an area that will play a major role in natural products chemistry during the next century. The past five years have already seen a rapidly developing interest in the chemistry of marine microorganisms and this has been reflected by the increasing number of literature reports related to microbial metabolites.

Genetic engineering is also predicted to have a large impact on the study of marine natural products. With the possibly of biosynthetic gene transfer from one organism to another, it appears that the fermentation of genetically modified microbes will be a future source of unique and highly desirable metabolites. This is especially relevant for the biomedical industries growing interest in marine natural products chemistry and the likelihood of a marine-derived drug reaching the marketplace in the near future. Genetic engineering, fermentation and aquaculture methods will have to be further developed in order to provide marine organisms for commercial production, since the total synthesis of particularly complex bioactive compounds will not always be possible or practical. This advancement in biotechnology may hopefully provide an acceptable means for supplying marine natural products, while at the same time preventing excess harvesting of the fragile marine environment.

The UNCLOS address neither marine genetic resources, nor bio-prospecting but only 'marine scientific research', where non-commercial purpose is supposed. The CBD just considers genetic resources as components of biological diversity. It is the sole international framework regulating the access to genetic resources. Its Article 15 states that national governments determine their own legislation to provide access to their own genetic resources—i.e. the ones included within their EEZ.

The CBD entered into force on December 1993, however, as the development of new techniques for exploitation, extraction and uses of genetic materials found at deep seabed-vent sites made its advances in the 1990s, it fails to adequately address these marine resources. Presently, although for the continental shelf non-living resources there are various agencies in charge of regulating most aspects of exploitation, for companies interested in bioprospecting no one appears to be in charge.

With regards to genetic resources within the *Area,* no regulation has been stated up till now. It has been noticed (Glowka 1996) that, giving the advances in underwater technology and biotechnology, and their present free accessibility, genetic marine resources may be the most immediately exploitable and lucrative ocean resource. Pharmaceutical research and development projects are increasingly carried out through international partnerships among diverse organizations to take advantage of differential expertise, technology and access to biological materials. Regarding investment in this field, it is worth noting that pharmaceutical companies do not fund discovery, they only fund activities after a useful agent has been found. The organism and the attached genetic material acquire economical value only after the additional considerable investment in research is made by the industry looking for biotechnological applications. Since genetic materials can now be reproduced in laboratories, there is no threat for species over-exploitation, depletion and conservation. Thus, in principle, genetic resources are infinite and non-rival in consumption, given the possession of the necessary technology to harvest and reproduce them. Bioprospecting firms take from these resources only their genetic information. Thus their regulation, if needed, should be different from the ones regarding fishing or mining.

Marine biotechnology needs a global organizational effort to achieve sustainable socio-economic benefits for mankind. In 1989 the International Marine Biotechnology Conference was established as an international forum to exchange information. It was held in Japan (1989), the United States (1991), Norway (1994), Italy (1997), Australia (2000), Japan (2003) and Canada (2005). Other international regional congresses—such as The Asia Pacific Marine Biotechnology Conference—and associations—such as The European Society for Marine Biotechnology—have been created as well.

However, to implement the commercialization of marine biotechnology products requires a strong partnership between scientists, innovative companies and also governments, since mechanisms to simplify the process that brings medical applications from the research stage to the market are needed. Governments must increase their involvement in marine biotechnological research facilitating the product pipeline and protecting intellectual property rights. Protection of intellectual property rights for potential products through patents, copyrights, trade secrets, etc is also fundamental for encouraging commercial development. Also, with reference to the area, a regulation protecting the 'genetic information right' (pertaining to the international community) should be developed in parallel with the intellectual property right protected by the patent.

Conclusion

This book has examined present and potential exploitation of marine natural resources from three different perspectives: the economic; the technological; and the environmental. In an effort to understand the problems at stake when exploiting natural resources from the sea, a strong emphasis has been put on the technological apparatus and the consequent environmental impact that it provokes on the marine ecosystem. The aim of this closing section is to summarize the findings for each of the five types of resources analyzed throughout the research: offshore oil and gas, marine renewable energy, underwater mineral deposits, fish, and marine bio-products.

An important contribution of this research is to describe state-of-the-art economics of marine natural resources in a global and comprehensive fashion. The economic viability of each exploitation process is a function of the technological capacity to extract each resource efficiently. Economic conditions and technological apparatus are explained in detail for each of the five marine natural resources examined in this research. The distinctive types of environmental problems that arise in the different exploitation processes are highlighted and possible solutions indicated. The basic common feature of each exploitation process is the uniqueness of the environmental characteristics in which it takes place. The complex dynamics that govern the ocean are described in chapter 1. The oceans play an essential role in the planet because they perform a number of fundamental ecosystem services. The oceans are the ultimate recyclers of wastes, they are the main determinant of the planet's climate, and they sustain the largest part of the planet's biodiversity. The oceans' elements and functions are interconnected in dynamic non-linear processes and interact at different levels.

As noted in chapter 2, human understanding of the dynamics of the marine ecosystem is still very limited. The causes of important regional and global climatic phenomena are not completely understood. This is relevant at various levels, particularly because it is impossible to precisely forecast natural catastrophes. The functioning of many oceanic mechanisms remains highly indeterminate and the environmental effects of large-scale economic activity are definitely unpredictable. Consequently, when dealing with oceans governance, it is impossible to conveniently separate naturally occurring phenomena from the effects of human activities. Moreover, as noted in chapter 1, the chain of effects and counter-effects is so integrated that it is not possible to cope with marine activities separately. The result of such interrelations is the global and interdisciplinary approach to oceanic problems, also called the ecosystem approach.

Another common feature emerging from this research is that information on natural marine resources is difficult to gather, expensive, and time-consuming. As noted in chapter 2, today marine research and exploration employ optic and acoustic sensors and modern research platforms and facilities. Marine technology has frequently originated from military research for the ultimate benefit of the whole scientific community. The relatively recent possibility to acquire information remotely has increased quality and quantity of data. Data is processed electronically with a tremendous saving of time and increased reliability. Oceanographic research instrumentation has improved the accuracy of ocean scientific surveying, definitely providing a more reliable and detailed information set.

With the ocean possessing the largest ecosystem, understanding the functioning of its dynamics is extremely important to be able to apply scientific knowledge to manage the planet and exploit its natural resources in an efficient and sustainable way. The more that is understood about global marine processes and linkages, the easier it is to monitor and predict the environmental effects of human activities and to distinguish them from changes caused by natural impacts. However, it is presently impossible to predict and manage effectively the impact of economic activities—especially in the long run—because the quality of information on many ocean and global processes is still too poor. Consequently, ocean governance is still heavily characterized by decisions in condition of uncertainty.

From the economic point of view, it has been seen that the ocean ecosystem supports three main economic functions: Resource Supplier; Waste Assimilator; and Direct Source of Utility. The focus of this research is on the first function. The ocean is a compound asset that provides a wide variety of goods, services, and energy. The first chapter has provided reasons why assessing the monetary value of the oceans is an almost impossible task. It has been found that traditional economic theory does not provide adequate instruments to measure the real economic value of the oceans' stock of goods and services, because the oceans are an essential good. However, it is possible to quantify the value of the flow of goods and services from the given stock of resources—i.e. the monetary value added by the harvesting production.

Theory and data presented in chapter 1 demonstrate the fundamental influence of the ocean on the global economy regardless of any quantitative attempt to put a monetary value on the ocean and its ecosystems. The ocean provides a number of basic ecosystem functions, and supplies goods and services that, in turn, generate a number of marine-related satellite industries having a considerable impact on the world economy. Then again, the focus of this book is on the ocean as a provider of natural resources, which can be divided into: biological or living resources; non-biological or non-living resources; and bio-power energy resources. As noted in chapter 1, some resources are renewable, while others deplete with exploitation.

With regards to bio-power resources, the study of chapter 4 reveals that various forms of renewable energy from the ocean cannot be exploited yet because of lack of proper technology. For non-living resources, chapter's 5 study on underwater mineral deposits shows that some resources are not yet economically exploitable. The main reason is the high costs of the technological equipment needed to operate at depth. As seen in chapter 2, the technological apparatus to explore marine resources located in the deepest parts of the ocean is still in its infancy. Consequently, it has been assessed that some sectors of the ocean industry cannot yet compete with the lower costs of land-based production of similar commodities. However, findings of this research study suggest that they most likely will in the near future. Perhaps this statement may sound trivial and naïve to experts, as it has been repeated constantly for several decades but never realized. It is, in fact, difficult to predict *when* marine technology will undergo a real breakthrough in reducing its production costs. However, given the increasing and relentless trend of technological development, it is naïve to think it will not.

Another finding of this research is that the ocean is marked by strong asymmetric evaluations. The demand for environmental protection and conservation is highly income-elastic. This implies that clean oceans and biological diversity are more valued by rich countries than poor ones. Besides, environmental problems are usually perceived with different time scale perceptions between cause and effect, both in different countries and among experts. The time-scale vision of economics and ecology is different as well. Economic projects, even in the long run, have a much shorter timeframe compared to ecologic processes. Another asymmetry concerns the different degree of development of monitoring and production technologies. Understanding of the marine environment and its dynamics is minor compared to the capacity to alter it through exploitation. The general recommendation is that the economic dimension of development should integrate the environmental dimension of protection and conservation. However, even if almost anyone agrees that economic growth should match the environmental protection requirements, the present hyper-intensive use of the ocean is without precedent.

As found throughout chapters 3, 5, and 6, living and non-living resources are being threatened as never before. Human exploitation adversely impacts, changes, and even destroys natural ecosystems. As confirmed throughout the whole research, marine ecosystems are increasingly affected by a variety of human and natural impacts. It has been noted that this trend has already gained momentum, and even if measures are taken immediately it may be possible to succeed in slowing the tendency but not in reversing its direction in the short run. Data reported in chapter 6 indicates that the decline of ocean wildlife is evident and that the increasing pressure of human overexploitation is seriously threatening the very existence of the ocean's living resources. It is only by maintaining and improving the environmental quality of the ocean that the value of its resources can be maximized. In chapters 1 and 3, it is argued that economic activity cannot stop just for the sake of conserving the environment. However, at the same time, everything possible to reduce the magnitude of the environmental impact must be done.

Chapter 2 clearly indicates that the ocean inherently needs global scientific research and governance. Effective coordination of national and international policies at various levels is needed in order to conserve and protect the marine environment. Policy makers and implementing bodies should be supported by reliable and updated scientific knowledge of ocean processes, as well as sound assessments on the status of marine resources. Scientific information-based decisions would contribute to shape appropriate and suitable industrial policies at both regional and local levels. A proper large-scale approach to ocean management has to be built on several international levels stemming from a thorough and well-founded diagnosis of the problems. Reliable and widespread scientific information would assist in managing effectively and efficiently local marine resources and coastal areas. However, as noted in chapter 2, the great amount of information required for sound management of marine resources is expensive to gather and not easy to interpret.

Ocean research does have a tradition of international cooperation and partnership, which is necessitated by the nature and vastness of the oceanic environment. However, as noted in chapter 2, the exploration of the ocean requires sustained effort with wide international participation. No nation has the intellectual, financial, and technological capabilities to undertake a thorough program of global ocean exploration alone.

As stated in chapter 1, the ocean is a common asset. This implies a number of inefficiencies because the market mechanisms do not work. Ocean governance is also plagued by transnational externalities. Given the global common nature of the ocean, the related problems involve nearly all nations of the world to be solved. This implies the adoption of international agreements at global scale, balancing different and often controversial interests and preferences, which usually reflect different stages of economic development, culture, and political structure.

Chapter 1 suggests that eventual solutions to these paramount problems rest on difficult and time-consuming international negotiations and agreements, which are hardly ever enforced. To accomplish global ecosystem-sound marine governance and management, a worldwide accepted and powerful global agency is suggested.

The main practical problem with the marine environment is that water is much denser than air. Consequently, it is much more difficult to access and operate. The harsh, turbulent and corrosive conditions at sea and the huge pressure at depth make the technology required for marine research and exploitation tricky and expensive.

Water and underwater marine technology have civilian applications in both scientific research and offshore economic activities. As discussed in chapters 2 and 3, the oil and gas industry is presently one of the most technologically advanced industries in the world. It is the locomotive of the entire marine civilian industry. Most underwater technological innovations in marine research and commercial sectors spin off from the offshore oil and gas sector. The offshore oil and gas industry has also a considerable positive impact on other sectors of the economy.

The importance of the offshore oil and gas industry in the world economy has been pinpointed in chapter 3. Offshore oil and gas productions contribute respectively around 35 percent and 25 percent of the world's supply. Since the first successful commercial venture in the mid-1950s, offshore petroleum production has constantly increased. Today more than 8,300 offshore oil and gas installations are operational worldwide. The success of offshore oil production has occurred despite the extremely large investment required for its development. The production activity has moved gradually further out offshore into deeper waters and a harsher operating environment.

As data given in chapter 3 demonstrates, even though improvements in technology and management practices could help reduce the magnitude of the environmental impact of the offshore oil and gas exploitation, there are negative effects on the marine ecosystem at all phases of the industrial activity. However, given its importance for the world economy, hydrocarbon exploitation cannot stop because of its negative environmental effects or the risk of possible accidents. The benefits coming from oil and gas production are much larger than the environmental damage caused by the necessary operations. Until new substitutes are found or renewable forms of energy developed extensively, hydrocarbon production must increase to sustain and fuel the growing world economy.

Chapter 4 has stressed that conventional forms of energy provoke pollution that damages the environment, its natural resources, and the health of its inhabitants. The increasing international agreement on the risk related to the production of greenhouse gasses and the threatening effect they have on global warming, climate change, and other environmental dynamics have been emphasized. However, it has been found that the pollution/health issue is just one of the main reasons why industrialized countries are increasing their efforts to shift towards more environmentally friendly forms of energy. The other reasons are the scarcity of conventional energy resources and the dependence on imported energy. These reasons have fuelled interest to develop different renewable energy systems, including marine energy converters. This research found that renewable energy sources are still expensive to harness with the present technological apparatus and that they are not commercially competitive with conventional energy. However, conventional energy production does not include its high environmental cost, while renewable energy plants have very low external costs. A socially and environmentally oriented comparison should, therefore, include external costs in the account.

The oceans contain a quantity of renewable energy that is greater than the present world's electrical energy consumption. However, even if ocean energy converters have been tested quite extensively and some pre-commercial demonstrations are ongoing all around the world, compared to other sources of renewable energy, ocean technology is still at the 'invention and research' stage. In fact, as chapter 4 demonstrates, the technology for extracting energy from the oceans is still crude and prototypes abound because of the immaturity of the sector. Being a promising industry still in its infancy, the renewable energy sector needs a supportive political and legal framework to help its development and an eventual breakthrough to be found.

The ocean is expected to provide in the near future an efficient, reliable, cost-effective, and environmentally friendly source of electrical power. Compared to other renewable forms of energy such as solar and wind power, which rely much more on atmospheric conditions to work, marine energy is more stable and therefore more reliable. The main environmental issues with offshore energy conversion facilities are associated with destruction and disturbance of the resident marine life and with degradation of scenic ocean front views. Another possible minor disturbance is the threat to navigation.

The ocean also contains unknown quantities of many different minerals, the potential value of which is indefinite. Even though the ocean crust and the continental crust are different, numerous processes of mineral formation are common on land and at sea. As speculated in chapter 5, it is expected that the average mineral content for each square kilometre of the seafloor is approximately equal to the land's one. Given the ocean's dimension, it is therefore evident that its deposits and placers have a far bigger mineral content than that of land. Some mineral commodities are already extracted from shallow water placers with profit. However, compared to its potential, present ocean mining is minor.

Discoveries of seabed mineralizations over time have generated interest in mining the oceans, arousing optimistic expectations among investors and pressing the international community to approve the United Nation Conference on the Law of the Sea. However, expectations were not fulfilled, and although throughout the years the apparatus to extract minerals from the oceans has improved significantly, today marine mining is technically feasible but not economically viable. The main reason for this is the high cost of production compared to land-based mineral extraction. This is due to the huge depth of the deposits, which are located at 3,000–5,000 m. Another reason is that there are still inland reserves that are not mined because the demand for some minerals is small. Therefore, uncertainty, high risk, and high costs still influence and prevent marine mining investment decisions.

However, as chapter 5 showed, continuous scientific discoveries and technological advances in mining give hope to realize this long-lasting dream in the near future. Probably offshore oil companies will dominate the ocean mining sector. They have a better chance to exploit seabed minerals because of their technological capability to operate at huge depths. Equipment they already possess can be modified or transformed to suit mining operations.

A critical issue in mining the sea is the potential threat to the marine environment. Mining has destructive effects on the seafloor and the benthic communities, and considerable effects on the water column. Obviously, the magnitude of the impact depends upon the technology employed. Moreover, marine mining can conflict with the exploitation of other marine resources such as biogenetic materials.

A recurrent theme of this research is the scarce knowledge of the oceans and the consequent urge to learn more about this grand reservoir of natural resources. In particular, as chapter 6 has showed, when living creatures are concerned, scientific uncertainty dominates the scene due to lack of reliable data. The reason for this gap is the vastness and density of the oceans compared to the technological apparatus available to marine researchers. Oceanic biological research is difficult, time consuming, and involves high costs. Therefore, long-term prediction is almost impossible.

From a market perspective, seafood provides high-quality proteins for human consumption and income for more than 100 million workers worldwide. It has been showed that from an economic viewpoint capture fishery production is inefficient because of the open-access common nature of the ocean living resources. Subsidy policies bring with them even more inefficiency in the fishery arena. As fish are public assets, fishermen should pay a price to the state. The big paradox in fishery management is that not only do states not charge fishermen for what they gain from public resources, but reward them with large subsidies—estimates of which vary greatly between USD 15–55 billion worldwide.

The result of years of bad governance in this sector is evident in the effects provoked by over-fishing and harmful fishing. If this trend is not reversed immediately, within a few decades wild fish will disappear. Since human actions determine, over time, the harvesting rate provoking intertemporal effects, in order to conserve wild fish stocks an efficient sustainable level of harvesting must be determined and enforced. Due to the lack of necessary data required to assess sustainable fish harvesting, the precautionary approach has been introduced at all levels by FAO's Agreements and Codes in order to conserve and manage living stocks. More recently, the ecosystem approach has been introduced. A complex mixture of biological, social, economic, and environmental information affect the decision-making process in fisheries management. However, the main difficulty is to enforce policies. Fish stocks are overexploited and depleted, by-catch figures are shamefully high, and harmful fishing still a diffuse practice that provokes destruction of marine habitats, ecosystems, and biodiversity.

Worldwide demand for seafood is increasing but faces a supply of wild fish characterized by a downward trend. On the other hand, there is increasing production of cultured fish, which is developing at a two-digit annual rate of growth. Today aquaculture accounts for over 30 percent of the seafood production, and will probably substitute completely capture fisheries within a few decades. Aquaculture supplies high demand species in a more stable fashion than capture fisheries and at lower price.

Chapter 7 presented an analysis of natural product derivatives in the drug industry, stressing its great importance for mankind. In terms of evolution and biodiversity the sea is far superior to the terrestrial ecosystem. During the evolutionary process, numerous forms of marine life have developed unique and complex biological mechanisms capable of synthesizing and releasing chemical agents.

The oceans started to attract the interest of pharmaceutical companies and research institutions approximately 50 years ago. Thousands of marine metabolites have been isolated, many of which are structurally novel and unprecedented within the terrestrial biosphere. Some marine organisms contain unique natural toxic chemicals that might also attack human cancer cells. New biochemical substances derived from marine creatures have already reached clinical trials.

Many marine-derived products other than drugs are also already on the market. The past two decades have seen a blossoming of applications of marine-derived products in cosmetics, nutraceuticals, and enzymes, each of which with a potential multi-million dollar market value. Chapter 7's findings underline that although still very few marine-derived pharmaceuticals have become marketable drugs, there is great potential that many marine products will eventually do so. The increasing number of patents in recent years is evidence of the growing industrial significance of marine natural products.

Even though the marine bio-product sector is not yet commercially developed, there are sound reasons to believe that it will be in the next few decades. Given its biological and chemical diversity, the marine environment has an enormous wealth in terms of bio-products. Marine genetic resources can have numerous applications in pharmaceutical, agrichemical, environmental, and commercial sectors. Advances in diving technology are opening new deep underwater territories to bio-prospecting. However, marine biotechnology needs a global organizational effort to achieve sustainable socio-economic benefits for mankind. Governments must increase their involvement in marine biotechnological research, at least facilitating the product's pipeline and effectively protecting intellectual property rights.

One of the central themes that run throughout this book is the issue of sustainability. Human ingenuity and consequent technological developments have enhanced economic growth and improved the standard of living for billions of people. While each chapter tells its own story, the lesson collectively learned through this research is that the pattern of increasing consumption has fuelled the exploitation of natural resources with little regard for consequences to the environment. The increasing rate of exploitation of natural resources is reducing the value of the natural capital. Facts and figures presented in this research strongly support the hypothesis that present technology and practices are not sustainable for the environment.

The natural resources' harvesting rate is a function of both population and technology. An increasing population needs a greater amount of resources to be sustained. If the exploitation process were going to continue with the present technology, these would result in scarcity of resources with consequent increase in prices. This, in turn, would widen the consumption gap between rich and poor. In this perspective, the role of technology is crucial. If the increasing population were going to use the same old and polluting technology in the years to come, the only possible outcome would be depletion of resources. On the other hand, environmentally friendly technological advances based on renewable sources can sustain a growing population. However, sustainable development is a difficult objective and there is no simple prescription for implementing it.

One issue that looks crucial is to understand ecosystem mechanisms and dynamics. There are several critical areas that need far more research. Large-scale collaborative studies of the marine realm are needed to enhance the understanding of how interrelated variables affect final outcomes. Scientific research expands knowledge of the environment and assists to manage it efficiently, conserving the natural capital without destruction and waste of natural resources. Another challenge ahead is to develop efficient ways to exploit renewable energy, especially from the sea because of its enormous potential. In this field, it is imperative that what is now technologically possible becomes technologically economical. In chapter 4 it has been noted that the ocean is a huge source of renewable energy. Governments' commitment to renewable energy from the oceans must be sustained over the long term or the nascent market will not develop.

On the policy front, as the market does not take into account indirect costs, such as environmental ones, income tax should be reduced and tax on environmentally destructive activities increased. In this way the market can incorporate external costs, and prices will better reflect the full costs of goods and services, including the ecological aspect. An economy is sustainable only if it respects the basic principles of ecology. Inadequate funding for activities such as monitoring reduces the ability to anticipate environmental problems. As a result, by the time a negative trend is observed it may be too late to halt it. Governments generally react to a risky situation only after it has reached the crisis point. Capture fisheries is a dramatic example of this practice.

Sustainable harvesting of natural resources requires changes in management approaches that emphasize precaution and imply minimization of ecological impacts and long-term biological conservation. Subsidies to fishermen should be abolished. There is growing awareness of the risks associated with mariculture (e.g. introduction of diseases and invasive species, dilution of gene pools, and increased biological interactions with other species). Sustainable mariculture can be achieved with responsible application of technology and the use of indigenous species. Moreover, regulation in this field is urgently needed. In the meantime, cultured fish can satisfy the needs of the increasing world population and release pressure on wild stocks.

Notes

CHAPTER 1

1. An ecosystem can be defined as a dynamic complex of interactions among organisms, such as animal, plant and microbes, and physical environmental features of a certain area.
2. The climate, in turn, influences hydrological and sedimentary cycles and determines the temperature of the upper levels of the water column, causing changes in marine ecosystems and related food chains.
3. Estimates of the quantity of annual oceanic absorption of CO_2 vary broadly. This quantity is the sum of two fluxes: the natural carbon cycle and the anthropogenic (human) induced carbon cycle.
4. *Plankton* is a general term to describe small organisms (plants, animals and algae) that passively drift in the sea. The *phytoplankton* is the smallest one, a single-celled organism with a length of about 0.0002 cm. It contains chlorophyll that allows it to carry out photosynthesis. It is at the bottom of the oceanic food chain and lives mostly in the top 100 m of the sea, where sunlight penetrates.
5. The present methods for measuring CO_2 were developed in the 1960s. A conventional measurement is the partial pressure of carbon dioxide (pCO_2). The flux of CO_2 between the atmosphere and the ocean is highly variable in time and space. Temperature is the most significant factor and warmer waters hold less CO_2. Under some circumstances the direction of the flux can invert.
6. A large marine ecosystem is a large region of the ocean (> 200,000 km^2) with unique bathymetry, hydrography and biological productivity.
7. Public goods are special goods that have two key characteristics: (1) they are non-excludable and (2) non-rival in consumption. When we say that a good is non-excludable, we mean that once the good is provided for one person it is either logistically impossible or extremely expensive to exclude other people from consuming that good. Typical examples of public goods are national defence, street lighting and lighthouses. A good or service is non-rival if the consumption of one person does not affect the quantity available for consumption by others. Non-rivalry means that the marginal cost of supplying the good to an additional user is zero. Once the good is provided for one person, it is available to all.
8. Water is scientifically defined as a mineral. Freshwater by desalination is an energy intensive process. However, as the world's need for freshwater is increasing, this industry is expected to grow. Water by desalination is actually produced in Algeria, Bahrain, Egypt, Iran, Israel, Kuwait, Libya, Malta,

Mexico, Netherlands, Oman, Qatar, Russia, Saudi Arabia, UAE and in many Caribbean countries.

9. One-third of the word's supply of Sodium Chloride is evaporated from seawater held in ponds.
10. For integrated coastal zone management see the Global Programme of Action for the Protection of the Marine Environment from Land-Based Activities (1995), UNEP(OCA)/LBA/IG.2/7.
11. The United Nations defines sustainable development as "Development that meets the needs of the present without compromising the ability of future generations to meet their own needs." (*Brundtland Report*, WCED 1987). See the United Nations website: http://www.un.org/esa/sustdev/sdissues/sdissues.htm (accessed on June 17, 2005).
12. An isobath is an imaginary line passing through all points of equal depth.
13. The ISA was established in 1994 and it is headquartered in Kingston, Jamaica.
14. UNCLOS is only legally binding for the states that signed for it. A relevant point is that the United States is not a signatory to UNCLOS.
15. A free-riding activity is a situation that commonly arises in public goods contexts. An individual (the free rider) may benefit from the actions of others without contributing. Each person has an incentive to make others pay for the provision of a public good without personally contributing.
16. An efficient property rights structure has three main characteristics: Exclusivity; Transferability; and Enforceability.
17. An equilibrium is Pareto-efficient if it is not possible to make anyone better off without at the same time making someone else worse off.
18. For a complete list of the ocean-related Organization see: Edgar Gold and Christopher Petrie, "International Maritime Organizations and Institutions", in *Maritime Affairs: A World Handbook* (1991) Longman Group UK Limited.
19. The precautionary principle means that in face of uncertainty about potentially irreversible environmental impacts, decisions concerning the use of environmental resources should err on the side of caution. It was declared in the UNCED that lack of scientific certainty is "no reason to postpone action to avoid potentially serious or irreversible harm to the environment".
20. The polluter pays principle requires that polluters must pay the cost of the environmental damage they provoke, either by paying the cost of monitoring and managing properly their activities or the cost of clean-up actions.
21. Preventative action is a proactive process intended to prevent potential concerns before they occur or become severe.

CHAPTER 2

1. On 23 January 1960, Mr. Jacques Piccard and US Navy Lieutenant Donald Walsh descended aboard the Swiss built bathyscaphe *Trieste* to the Challenger Deep in the Mariana Trench, at 10,916 m depth.
2. Co-founder of the computer components company Intel, Gordon Moore in 1965 stated that the number of transistors on a computer chip doubles about every two years.
3. The bathythermograph was developed to locate thermoclines in order to detect submarines.

4. In addition to direct warfare support, BT observations are also used to analyze location and structure of ocean's fronts and eddies, predict ocean currents, surface temperature and other physical features.
5. The hydrophone is a passive SONAR. It converts acoustic energy into electrical energy and is used in passive underwater listening.
6. Since the 1970s, there are some regional pollution monitoring programs operational in the Baltic Sea, Northeast Atlantic/North Sea and the Mediterranean. A wider program for monitoring the oceans is the Global Oceanic Observing System (GOOS), which—since 1998—works together with the GTOS (Global Terrestrial Observing System) and the GCOS (Global Climate Observing System) as part of IGOS (Integrated Global Observing Strategy). There are also numerous international organizations developing indicators specifically related to the oceans, among them the Commission on Sustainable Development (CSD), the Group of Experts on the Scientific Aspects of Marine Environmental Protection (GESAMP) and the World Resources Institute (WRI).
7. The presence of mineral resources on the seabed and in the subsoil depends on the geological stage of the ocean area. Tectonic plate movements and dynamics are believed to be the main long-term cause of volcanism, which in turn is responsible for the presence and diffusion of mineral formations and deposits at sea.
8. Dr. Jack Corliss.
9. See chapter 7 for a description of these organisms.
10. A transducer is a device that converts energy (sound, temperature, pressure, light or other types of energy) from one form to another. Even common microphones and light bulbs are transducers.
11. Doppler effect is the shift in frequency between the signal transmitted and the echo received back to the acoustic sensor. It is the same principle of fluorescence in optics.
12. A short analysis of produced water is given in the chapter on offshore oil and gas.
13. Excess of nutrients such as nitrogen and phosphorous (coming from discharges of human activities) can alter the marine ecosystem. Starting from an increase in *phytoplancton* production, excessive nutrients at sea provoke an unbalance in the relationships among species. An unbalanced ecosystem may experience oxygen depletion (Hypoxia) or produce phenomena such as algal blooms and coral bleaching.
14. Usually there is a delay between the time the sensor detects the target and the time the data is available for use. If this delay is short (few hours) the data is said to be "near real time."
15. For a complete specification of fisheries acoustic surveying see the NOAA 2004 publication on line at: http://www.st.nmfs.gov/st4/protocol/acoustic_protocols.pdf (accessed on March 10, 2006).
16. Underwater operations in the offshore oil and gas sector include tasks such as marine riser deployment, vessel/rig selection, ROV operations, station holding, ship loading etc.
17. The Ocean Information Center at the University of Delaware (USA), which provides figures on voluntary information submittals, estimates that the world's research fleet consists of nearly 500 vessels and about 60 AUVs from 53 nations.
18. Airborne platforms such as piloted, remotely controlled and autonomous aircrafts are an integral part of modern marine research.
19. http://www.whoi.edu/ (accessed on 15, January 2006).

20. One of the latest applications of manned submersibles is tourist submersible cruising, which is an increasing market with an operational fleet of about 50 units. Submersibles can host 30–60 passengers for hourly excursions at depths of about 50 m.
21. Submersible *Alvin*, operated by the Woods Hole Oceanographic Institution in Massachusetts, discovered the wreck of HMS *Titanic* in 1985.
22. A payload is any piece of equipment that performs a function that is not essential to the guidance, propulsion or control of the vehicle. However, it is of fundamental importance in order to achieve the mission target.
23. http://www.dsl.whoi.edu/ (accessed on 15 January, 2006).
24. Lithium-based batteries have 10 times the energy density than manganese alkaline ones, but are about 10 times more expensive.
25. Fuel cells convert energy stored in fuel directly into electricity not by burning the fuel, but through an electromechanical process.
26. GPS, originally intended for US military use, is funded and operated by the US Department of Defense. The system is made up of a set of 24 or more satellites, each of which orbits the Earth once every 12 hours. Signals from at least four individual satellites are necessary in order to calculate a position in three dimensions, together with an exact time signal. For some years now, signal access has been available worldwide to civilians just with the use of a commercial GPS receiver.
27. To obtain high-resolution maps of the seafloor, high-frequency sound is required. Since high-frequency sound is rapidly attenuated by seawater, to produce good-quality maps it is necessary to operate close to the bottom. The traditional method to map the seafloor employs a R/V towing SONAR device close to the seabed. For deep-sea rough-feature seafloor mapping, the traditional method can be slow and dangerous (the device can be damaged). Therefore, AUVs are now employed in this activity, since they provide a faster and a more stable platform to perform the same task more efficiently. AUVs used for mapping are usually equipped with a payload that includes multi-beam SONAR for swath bathimetry, one or two side-scan SONAR producing high-resolution imagery and a sub-bottom profiler allowing the detection of sub-bottom features.
28. Sensors can detect the distress signal emitted by vessels in difficulty. The recorded signal is used to locate the position of the vessel.
29. Ocean observatory NEPTUNE is placed at Juan de Fuca tectonic plate, off the North American West Coast. When completed, NEPTUNE is supposed to be able to monitor more area of the ocean and seafloor than any other observatory. More than 3,000 Km of fibre optic cables will connect the numerous sensors and stations of the network.
30. http://www.jamstec.go.jp/jamstec/station.html (accessed on 17 January, 2006).
31. Technical information about drifters can be found at http://www.technocean.com/ (accessed on July 14, 2006).
32. *Spray* was the name of the sailing boat of the famous navigator and solo sailor Joshua Slocum.

CHAPTER 3

1. Crude oil—or petroleum—includes condensate (from separator on wellhead), natural gas liquids (NGL from gas processing plants), synthetic oil and refinery processing gains, other oils and stock withdrawal.

2. Apart from emission problems, this is another reason why oil should not be burnt.
3. The standard unit of measurement for all oil production and sales is the barrel. One barrel equals 159 liters.
4. Unconventional sources are heavy oil and tar sands, not yet economically exploitable with current technology.
5. http://www.worldoil.com (accessed on May 13, 2006).
6. http://www.bp.com (accessed on May 13, 2006).
7. http://www.ogj.com (accessed on May 13, 2006).
8. http://www.energy.ihs.com (accessed on May 13, 2006).
9. http://www.exxonmobil.com (accessed on May 13, 2006).
10. http:// www.usgs.gov (accessed on May 13, 2006).
11. Other major breakthroughs in offshore oil technology can be traced back to 1953—when the first movable offshore drilling rig, called 'Mr. Charlie' was built in the Gulf of Mexico—and in 1962, when the first semi-submersible drilling rig was launched by Shell.
12. For a complete list of the projects' locations, see the website http://www.offshore-technology.com (accessed on May 18, 2006).
13. http://www.spe.org (accessed on May 18, 2006).
14. The concept of "deepwater" changes with time. Four years ago the threshold was a depth of more than 200 m, now it is more than 500 m.
15. In the same year ultra-deepwater oil production was about 3 percent of total world production.
16. A piezoelectric devise can generate a voltage when a mechanical force is applied.
17. Produced waters include formation water, brine, injection water and other technological waters. Formation water and brine are extracted along with oil and gas, as oil and gas reservoirs have a natural water layer that, being denser, lies under the hydrocarbons. Oil reservoirs often contain large volumes of water, while gas reservoirs only small quantities. To achieve maximum oil recovery, additional water (injection water) is usually pumped into the reservoirs in hundreds of thousands of tons to maintain the pressure in the system and force the oil to the surface. Both formation and injected water are extracted along with the hydrocarbons. As an oil field becomes depleted, the amount of produced water increases.
18. "Slurry injection" is the term that usually indicates the technique. However, other names are used as well, according to the company providing the service, such as: slurry fracture injection; fracture slurry injection; drilled cuttings injection; cuttings re-injection; grind and inject.
19. Signatory states of the OSPAR Convention are: Belgium, Denmark, Finland, France, Germany, Iceland, Ireland, Luxembourg, the Netherlands, Norway, Portugal, Spain, Sweden, Switzerland and the United Kingdom. The European Community is also signatory.
20. http://www.marinegroup.com (accessed on June 15, 2006).
21. Hydrocarbons break up and dissipate according to how persistent they are. Light products such as propane, benzene and kerosene tend to evaporate and dissipate quickly and naturally, they rarely need clean-up actions. These are called non-persistent oils. In contrast persistent oils, such as many crude oils, bitumens and resins break up and dissipate more slowly and usually require a quick clean-up response.
22. An emulsion is formed when two liquids combine, with one ending up suspended over the other.

23. Megapascal is a unit of measure for pressure. One Megapascal equals 10 bars (10 million Dynes per square centimeter), or approximately 145 PSI (pounds per square inch).
24. Discoveries of large gas hydrate accumulations had been made not only beneath the sea, but also in terrestrial permafrost regions of the Arctic.
25. Possible resources reported by the United States Geological Survey and others could amount to more than twice the energy equivalent of all other hydrocarbon fuels put together.

CHAPTER 4

1. Watt (W) is the yardstick for measuring power. A 100 W light bulb, for example, is rated to consume 100 Wh of power when turned on. If such a light bulb were on for five hours it would consume a total of 500 Watt-hours (Wh) of energy. W measures instantaneous power, while Wh measures the total amount of energy consumed over a certain period of time. For instance a 100 MW (1 MegaWatt = Million Watt) rated wind farm is capable of producing 100 MW during peak winds, but will produce much less than its rated power when winds are light. As a result of these varying wind speeds, over the course of a year a 100 MW power-rated wind farm may only average about 30 MW of power production. The ratio of a power plant's average production to its rated capability is known as *capacity factor*. The wind farm in this example would have a 30 percent capacity factor (30 MW of average production divided by 100 MW rated capability).
2. The main producers of wind energy are Germany, Spain and Denmark.
3. Actually there is also another possibility to harvest energy from the oceans, offered by the ocean salinity gradients. However, it is still a hypothesis.
4. An axial turbine is a turbine pitched at a 45° angle with the axis of the shaft.
5. In a direct driven generator—or power buoy—the buoy is the activated body. The buoy is directly connected to the translator located inside the generator through a rope. The motion of the waves, moving the buoy, is transmitted to the translator (Meyer and Nielsen 2000).
6. The Wells turbine was designed by Alan Wells in the late 1980s and is a special wind turbine capable of maintaining constant direction of revolution, whatever the direction of the airflow passing through is. Under the oscillating flow conditions of waves, Wells turbines have an efficiency of about 40 percent. Research is trying to develop variations on the Wells turbine in order to expand its performance.
7. http://www.wavege.co.uk (accessed August 15, 2006).
8. See: Clément A.H., P. McCullen, A. Falcão, A. Fiorentino, F. Gardner, K. Hammarlund, G. Lemonis, T. Lewis, K. Nielsen, S. Petroncini, M-T Pontes, P. Schild, B.iO. Sjöström, H. C. Sørensen and T. Thorpe (2002) "Wave Energy in Europe—Current Status and Perspectives", *Renewable and Sustainable Energy Reviews* 6: 405–431; and Electric Research Power Institute (EPRI) (2004) Assessment—Offshore Wave Energy Conversion Devices for further details about these technologies.
9. OSPREY was a 750-metric ton structure placed on the seabed 100 m from shore at a depth of 14 m in Dounreay, Scotland. Before the installation was completed, Osprey was torn apart by a storm.

10. LIMPET is the first commercial wave power station ever constructed. Wavegen, the same company that built OSPREY, implemented it. Given the previous experience, this time the converter device was placed on the shoreline.
11. http://www.oceanpd.com (accessed August 15, 2006).
12. The distance of the moon from the Earth varies cyclically with the moon's phases.
13. The head of a hydropower plant is the height of the standing water.
14. Another solution to handling a bi-directional flow is to turn individual rotor blades through 180 degrees at each slack tide.

CHAPTER 5

1. Phosphorite deposits occur on several continental shelves, such as the Atlantic waters of the United States, Morocco, Gabon, Congo, Namibia and South Africa and in the Pacific Basin in the United States, Mexico, Ecuador, Peru and New Zealand. Phosphorite deposits are usually found at water depth of less than 1,000 m. They are mined at sea and used at land mostly as agricultural fertilizer.
2. Rare Earth Elements is a series of 15 metallic elements used for heat resistant alloys. Among them Monazite and Basanite had been found at sea placers.
3. There is another a-biogenic theory about hydrocarbons genesis that is mentioned in chapter 3.
4. A placer deposit is an accumulation of mineral grains concentrated by sedimentary processes.
5. An ore is a concentration of minerals that can be mined, processed and marketed at a profit. Although the word 'ore' should only be used when a completed feasibility study has been done and the profitability of the investment assessed, in this work it is synonymous with deposit.
6. As the resources of the *Area* common heritage of mankind, this implies to share part of the revenue coming from the exploitation of these resources with less developed countries. This fact contributes to decrease the interest in mining the *Area*, shifting the interest to regions of the seafloor under national jurisdiction.
7. In order to improve the durability of the nourishment, the grain size of the borrow sand should be similar to the native sand.
8. Usually the criteria considered in evaluating license applications concern water depth, distance from the coast and depth of the extraction pit.
9. Some European countries such as the UK, the Netherlands, Denmark and France require a formal environmental assessment report of the impacts on beaches, coastal sediments, sand banks and bars, waves and currents and the existing marine flora and fauna.
10. Nodules average size is usually smaller, however, bigger nodules of 50 cm had been reported (ISA 2001b).
11. If the prices of the metals contained in the nodules rise or either the recovery cost falls, more nodules can be added as reserves.
12. The term "massive" refers to the metal content of these mineralizations, meaning that they have a high content of heavy metals.
13. On May 2006, Nautilus became a public company at the Canada Stock Market, raising more than USD 22 million.
14. Platinum Group Elements (PGEs) are: Platinum, Palladium, Rhodium, Ruthenium and Iridium.

15. http://www.iodp.org/ (accessed on February 11, 2006).
16. UNCLOS full text is available for download at http://www.oceanlaw.net/text/losc.htm.
17. On land, hundreds of holes are drilled to recover and analyze corers of the material in order to assess the shape, grade and tonnage of a mineral deposit.
18. See chapter 7 for an assessment of the scientific importance and economic value of the genetic material coming from these resources.

CHAPTER 6

1. Biomass is the quantity of a marine living resource as measured by kg tissue/m^2.
2. Taxonomy is the classification and identification of organisms. The traditional procedure to identify organisms involves comparing the physical characteristics of a collected specimen with the characteristics of a known species. There are numerous taxonomic books that describe the physical appearance, both externally and internally, of millions of species, as well as what it is known about their habitat and their general biology.
3. The main problem of classification is how to denote a species and hence the species concept. For example, a fish species consists of highly morphologically different males and females. Unless the fish are observed breeding, there is a real risk that they will be described as two distinct species. Thus, the so-called biological species concept of interbreeding is needed to establish them as one species. However, for the vast majority of small organisms nothing is known about their sexual reproduction—if they have any. Therefore, species are defined solely on the basis of their shape, the so-called typological or morphological species concept. This may both lead to both underestimation and overestimation.
4. The great productivity of coral reefs depends upon the productivity of the algae that cover the reef surface or inhabit the coral polyps themselves. The Great Barrier Reef off Australia hosts more than 300 species of corals, 1,500 species of fish and 4,000 species of molluscs.
5. Kelp is the popular name for a number of large brown algae that grow in the temperate waters along the rocky shorelines of North and South America, Europe, Asia, and Australia. Kelp forms dense forests from the sea floor to the surface, providing food and shelter for various kinds of marine plants and animals.
6. Upwelling areas are locations where cold, nutrient-rich water comes from the depth to the surface. As the surface water moves, it is replaced by water 'upwelling' from below. Due to the increase in *phytoplankton* growth (primary production), the most important fishing grounds can be found in upwelling areas. Upwelling areas only cover approximately 0.1 percent of the ocean surface, however, are responsible for about 25 percent of the world's fisheries.
7. http://www.mbari.org/microbial/ESP/ (accessed on November 20, 2006).
8. http://www.hafro.is/catag/ (accessed on March 23, 2006).
9. FAO reports that more than 50 species of fish and marine mammals are observable from low-level aircraft spotting (the application of remote sensing technology to marine fisheries, FAO Website, Fisheries Department).
10. In October 2004 OBIS recorded 40,000 marine species with 5 million records in its digital database. The information is available on the website. By 2007

OBIS is expected to develop 10 regional nodes to ensure and sustain global coverage and to serve at least 10 million records covering marine species and their location.

11. CeDAMar is carried out under the responsibility of the German Centre for Marine Biodiversity Research (Deutsches Zentrum für Marine Biodiversitätsforschung—Forschungsinstitut Senckenberg).
12. According to the FAO, fish provides over 7 percent of animal protein in North and Central America, more than 9 percent in Europe, over 17 percent in Africa, 26 percent in Asia, and 22 percent in low-income food deficit countries, including China.
13. Production of these species presently comes mainly from aquaculture, which has increased supply and reduced the price.
14. Chinese statistics had been challenged for inaccuracy by the FAO. Various studies have assessed that overestimation in Chinese statistics is about 50 percent. Further investigations are undergoing.
15. Fishmeal is created from cooking, pressing, drying, and milling of wild-caught, small pelagic fish such as anchovies and menhaden. About 5 kg of live fish is needed to produce 1 kg of fishmeal. Fish oil is usually a byproduct of the reduction process by which fishmeal is created (Delgado et al. 2003). The main use of fishmeal is for feeding terrestrial livestock such as poultry and pigs and farmed carnivorous fish such as shrimp and salmon.
16. World famous examples of failures to control overfishing include the collapses of: the Peruvian anchoveta (*Engraulis ringens* Jenyns) fishery in the 1960s and 1970s; North Sea herring (*Clupea harengus* L.) stocks in the late 1960s; the Canadian Atlantic groundfish fishery (where cod *Gadus morhua* L. was the target species) in the early 1990s; the Northwest Atlantic groundfish fishery (cod and haddock *Melanogrammus aeglefinus* L.); and the yellowtail flounder *Limanda ferruginea* (Storer) in the United States in mid 1990s.
17. Given the growing importance of aquaculture and the different problems and issues involved, it is useful to divide fisheries into two separate activities: capture fishery; and culture fishery.
18. The number of young fish that survive and eventually grow big enough to be caught by the fishery is called recruitment.
19. Total costs consist of fixed cost for fleet maintenance—which are independent of fleet utilization—and variable costs, which are directly related to the extent of fleet utilization. These are measured as a percentage of the fishing period in which the vessels are actually used for harvesting.
20. In a five-country survey conducted in Asia in the 1990s, an external team financed by FAO found that rates of return over variable costs in pond aquaculture were 150–200 percent (Delgado et al. 2003).
21. Many studies have shown that the local extent of altered benthic community structure and biomass is limited to less than 50 m.
22. Eutrophication is the process of natural or anthropogenic enrichment of water with inorganic nutrient elements.
23. Personal communication. Diver/instructor Antonio Anastasi reports that while diving under a tuna pen in Malta coastal water, a layer about 5 m thick of wastes was found.
24. In North Atlantic salmon farming, the number of escaped fish totals hundreds of thousands each year. In Norway, the main salmon producer, in 2005 about 732,000 escaped salmons were reported. It is likely that the true figures of escapes are much higher, since minor losses are usually not reported to the authorities. Many escapes are the result of net-pen failure and of boats

colliding with net-pens (Institute of Marine Research, *Challenges facing the coastal zone*, Norway 2006).

25. The term "mangrove" does not refer to a specific taxonomic group of species. One description implies all halophytic (salt tolerant) species of tropical trees and shrubs representing approximately 12 families and over 50 species, all adapted to living in loose wet soil, saline habitat, and periodic tidal submergence. Mangroves provide a number of ecosystem services, including nutrient filtration and cleansing, flood and storm protection and provision of habitat for many marine organisms useful to humans.
26. Some countries have posed specific regulations relating to aquaculture farm operation, such as effluent limits or design standards, or even requirements for social and environmental impact assessment (FAO 2001b).
27. The Bangkok Declaration can be downloaded at: http://www.enaca.org. (accessed on February 14, 2006)
28. The International Council for the Exploration of the Sea (ICES) is the organization that coordinates and promotes marine research in the North Atlantic area. ICES's fisheries' researchers sample a minimum of around 200 fish per 1,000 tons landed.
29. From logbooks it is possible to determine Catch-Per-Unit-Effort (CPUE) data. An increasing value of CPUE indicates a small availability of fish.
30. ICES's advice sets two minimum levels below which fish stocks should not be allowed to fall: (1) the biomass limit, and (2) the precautionary biomass limit. This second limit is posed to prevent the inherent uncertainty in estimates.
31. The 1995 Straddling Stock Agreement introduced this principle, aiming to prevent the absence of definitive and comprehensive information on the status of the stocks.
32. Indices of exploitation rates are obtained by comparing the indices of proven fishing stock to catches.
33. For instance, climate change is likely to effect the survival of fish eggs and larvae of cold-water species and may even alter the distribution of food at sea.
34. Regarding the world's fish stocks, the Johannesburg Declaration on Sustainable Development states that MSY should be achieved by 2015 and the ecosystem approach implemented by 2010.
35. Ecological and economic values of biodiversity are not necessarily the same. From an ecological point of view biodiversity may be important to the functioning of some ecological system without a market value. However, from an economic perspective the value of biodiversity is in its role in the production of goods and services. An individual harvested species is valued for specific properties that make it useful in either production or consumption. Since market prices influence the exploitation of a fishery, high effort toward particular species—fuelled by subsides—leads to the elimination of highly valued species and to reduction in biodiversity.
36. http://www.oceansaver.no/oceansaver/start.htm (accessed on November 11, 2006).
37. http://globallast.imo.org/index.asp?page=mepc.htm (accessed on November 13, 2006).

CHAPTER 7

1. Gorgonian is an octocoral (soft coral) having a flexible, often branching skeleton of horny material coming from the phylum cnidaria.

2. At about 150 m deep, the sunlight cannot penetrate the water and, consequently, photosynthesis stops. Chemosynthetic life means an ecosystem that thrives transforming chemical compounds in energy without the need of sunlight. Both photosynthesis and chemosynthesis involve an energy source (carbon dioxide) and water in order to produce sugars. Photosynthesis gives off oxygen gas as a byproduct, while chemosynthesis produces sulphide.
3. Phylogenetic is the study of the evolutionary relationships among groups of organisms.
4. *Hyperthermophiles* grow and dwell also in terrestrial geothermal areas such as geysers, hot springs and volcanic areas.
5. Proteomics is the study of the structure and functions of the proteins contained in the genome of an organism.
6. Presently TBT (tributyltin) is the predominant biocide in marine anti-fouling paints to prevent organisms from attaching to ship's hulls. TBT, which has been recognized as one of the most toxic substances ever introduced in the marine environment, accumulates in sediment and biota causing serious adverse effects in non-target organisms. The IMO banned TBT in 1990 for vessels not longer than 25 meters.
7. Although this strain was originally isolated from a hot spring in Yellowstone National Park, it also lives in hydrothermal vent sites.
8. For the latest list of published genomes see http://www.jgi.doe.gov (accessed on November 1, 2006).
9. Before a pharmaceutical company can initiate testing in humans, it must conduct extensive preclinical or laboratory research. This research typically involves years of experiments in animal and human cells. Compounds are also extensively tested in animals. If this stage of testing is successful, a pharmaceutical company begins testing the drug in humans. The clinical testing of experimental drugs in the EU and in the United States is normally done in three phases, each successive phase involving a larger number of people. Phase I studies are primarily concerned with assessing the drug's safety. This initial phase of testing in humans is done in a small number of healthy volunteers (20 to 100), who are usually paid for participating in the study. Once a drug has been shown to be safe, it must be tested for efficacy. This second phase of testing may last from several months to two years and involve up to several hundred patients. Most phase II studies are randomized trials. Only about one-third of experimental drugs successfully complete both phase I and phase II studies. In phase III study, a drug is tested in several hundred to several thousand patients. This large-scale testing provides the pharmaceutical company with a more thorough understanding of the drug's effectiveness, benefits and the range of possible adverse reactions. Most phase III studies are randomized and blinded trials. The first three phases constitute the clinical development and take from 5 to 10 years. Then the drug can be marketed.
10. Acyclovir, which was synthetically known as Ara A, was modelled based on sponge-derived spongothymidine or spongouridine. Ara A along with its acetyl congener Ara U, were later isolated as natural products from the gorgonian *Eunicella cavolini*.

Selected Bibliography

Ahlbrandt, A. (2000) United States Geological Survey: The USGS world oil and gas assessment. Paper presented at the AAPG Convention held in New Orleans, Louisiana, April 16–19, 2000. http://pubs.usgs.gov/dds/dds-060/ (accessed October 1, 2006).

Alldredge, A.L., D.L. Bradley, D. Butterworth and J.H. Steele (1999) 'Assessing the global distribution and abundance of marine life'. Summary of a workshop sponsored by the Sloan Foundation and the Office of Naval Research held in Monterey, California, January 13–15, 1998, *Oceanography* 12: 41–46.

Amador, M. L., J. Jimeno, L. Paz-Ares, H. Cortes-Funes and M. Hidalgo (2003) 'Progress in the development and acquisition of anticancer agents from marine sources'. *Annals of Oncology* 14: 1607–1615.

Angel, Martin V. (1997) *Pelagic biodiversity*. In: Ormond, R.F.G., J.D. Gage and M.V. Angel (eds.) *Marine biodiversity: patterns and processes*. Cambridge: Cambridge University Press.

Arif, J., M. Al-Hazzani, A. Amal, M. Kunhi and F. Al-Khodairy, (2004) 'Novel Marine Compounds: Anticancer or Genotoxic?' *Journal of Biomedicine and Biotechnology* 2: 93–98.

Bachmayer R., S. Humphris D. Fornari, C. Van Dover, J. Howland, A. Bowen, R. Elder, T. Crook, D. Gleason, W. Sellers and S. Lerner (1998) 'Oceanographic Research Using Remotely Operated Underwater Robotic Vehicles: Exploration of Hydrothermal Vent Sites on the Mid-Atlantic Ridge at 37° North 32° West'. *Marine Technology Society Journal*, 32 (3): 37–47.

Balmford, A., A. Bruner, P. Cooper, R. Costanza, S. Farber, R.E. Green, M. Jenkins, P. Jefferiss, V. Jessamy, J. Madden, K. Munro, N. Myers, S. Naeem, J. Paavola, M. Rayment, S. Rosendo, J. Roughgarden, K. Trumper and R.K. Turner (2002) 'Economic Reasons for Conserving Wild Nature'. *Science* 297: 950–953.

Bergstad, O.A. and O.R. Godoe (2003) 'The Pilot Project "Patterns and Processes of the Ecosystems of the Northern Mid-Atlantic": Aims, Strategy and Status'. *Oceanologica Acta* 25: 219–226.

Bhatt, V., A. Sarkar, R. Kumar, S. Basu and V.K. Agarwal (2004) 'Impact of Oceansat-I MSMR data on analyzed oceanic winds and wave predictions'. *Ocean Engineering* 31: 2283–2294.

Block, B.A., L.H. Teo Steven, A.Walli, A. Boustany, M.J.W. Stokesbury, C.J. Farwell, K. C. Weng, H. Dewar and T. D. Williams (2005) 'Electronic tagging and population structure of Atlantic bluefin tuna'. *Nature*, 434: 1121–1126.

Bockstael, Nancy, E. Freeman III, A.M. Kopp, R.J. Portney, R. Paul and K.V. Smith, (1998) '*On Valuing Nature*'. Manuscript.

Braden, John and C.D. Kolstad, (eds.) (1991) *Measuring the Demand for Environmental Quality*. Amsterdam: North Holland.

British Columbia Offshore Oil & Gas Technology Update (2001) JWEL Project No. BCV50229. Canada, October 19, 2001. http://www.offshoreoilandgas.gov.bc.ca/reports/jwl-report/intro.pdf (accessed August 27, 2006).

British Petroleum (BP) (2006) *Statistical Review of World Energy*. http://www.bp.com/liveassets/bp_internet/globalbp/globalbp_uk_english/publications/energy_reviews_2006/STAGING/local_assets/downloads/pdf/statistical_review_of_world_energy_full_report_2006.pdf (accessed September 21, 2006).

Brown, Lester R. (2001) *Eco-Economy*. New York: WW Norton & Company.

Busattil, S. (2004) 'Globalization and the Sea'. *Ocean Yearbook* 18: 117–122.

Census of Marine Life (COML) *Research Plan*, Version January 2005. Census of Marine Life International Secretariat. http://www.coml.org/baseline/coml_research_report_2005.pdf (accessed July 22, 2005).

Census of Marine Life-ChEss (2003) Biogeography of Deep-Water Chemosynthetic Ecosystems. Developing a long term field phase for understanding the biogeography of deep-water chemosynthetic ecosystems at the global scale. http://www.soc.soton.ac.uk/chess/chess_sci_plan.pdf (accessed February 15, 2006).

Center for Renewable Energy Sources (CRES) (2002) *Wave Energy Utilization in Europe. Current Status and Perspectives*. European Thematic Network on Wave Energy. http://www.waveenergy.dk/wave-rapport/2afp_rap/bku_afpr99/EU_WaveEnergy1.pdf (accessed April 24, 2006).

Churnside, James (2000) *Detecting Marine Life from the Air*, manuscript.

Cicin-Sain, B., R. W. Knecht, L. Denno Bouman, G. W. Fisk (1996) 'Emerging Policy Issues in the Development of Marine Biotechnology'. *Ocean Yearbook* 12: 179–206.

Clauss Gunther and Sven Hoog (2002) 'Deep Sea Challenges of Marine Technology and Oceanographic Engineering'. In Science-technology Synergy for Research in Marine Environment: Challenges for the XXI Century, London: Elsevier Science Ltd.

Coase, R. H. (1960) 'The Problem of Social Cost'. *Journal of Law and Economics*, 3: 1–44.

Collett, Timothy S. (2003) Natural Gas Hydrates: Resource of the 21st Century? Manuscript.

Committee on Exploration of the Seas, Ocean Studies Board Division on Earth and Life Studies, National Research Council of the National Academies (2003) *Exploration of the Seas: Voyage into the Unknown*. Washington, DC: The National Academies Press. http://www.nap.edu/catalog/10844.html (accessed June 1, 2005).

Committee on Marine Biotechnology. Biomedical Applications of Marine Natural Products, National Research Council (2002) Marine Biotechnology in the Twenty-First Century: Problems, Promise, and Products. Washington, DC: The National Academies Press.

Committee on Oil in the Sea (2003) *Inputs, Fates, and Effects*. Ocean Studies Board and Marine Board, Divisions of Earth and Life Studies and Transportation Research Board, National Research Council, Oil in the Sea III: Inputs, Fates, and Effects. Washington, DC: The National Academic Press, 2003.

Costanza, R., F. Andrade, P. Antunes, M. van den Belt, D. Boersma, D.F. Boesch, F. Catarino, S. Hanna, K. Limburg, B. Low, M. Molitor, J.G. Pereira, S. Rayner, R. Santos, J. Wilson and M. Young (1988) 'Principles for Sustainable Governance of the Oceans'. *Science* 281: 198–199.

Costanza, R., R. d'Arge, R. de Groot, S. Farber, M. Grasso, B. Hannon, K. Limburg, S. Naeem, R.V. O'Neill, J. Paruelo, R.G. Raskin, P. Sulton and M. van der Belt (1998) 'The Value of the World's Ecosystem Services and Natural Capital'. *Ecological Economics* 25: 3–15.

Crookes, W.J., L.-L. Ding, Q.L. Huang, J.R. Kimbell, J. Horwitz and M. McFall-Ngai (2004) 'Reflections: The Unusual Proteins of Squid Reflective Tissues'. *Science* 303: 235–238.

Cruickshank, M. J. (1998) 'Law of the Sea and Mineral Development'. *Ocean Yearbook* 13: 80–106.

Deere, C. (2001) 'International Trade, Conservation, and Sustainable Development in the Fisheries Sector: Conflict or Compatibility?' *Ocean Yearbook* 15: 102–220.

Delgado Christopher L., Nikolas Wada, Mark W. Rosegrant, Siet Meijer, and Ahmed Mahfuzuddin (2003) *Fish to 2020 Supply and Demand in Changing Global Markets*. Washington, DC: International Food Policy Research Institute and Malaysia: WorldFish Center Penang. http://www.ifpri.org/pubs/books/fish2020book.htm#download (accessed August 13, 2006).

Douglas-Westwood Associates (2000) 'UK Marine Industries World Export Market Potential'. A Report for the Foresight Marine Panel. Canterbury, UK: Douglas-Westwood Ltd.

Douglas-Westwood Associates (2004a) *Renewable Energy Key Facts & Figures 2004–2008*. Canterbury, UK: Douglas-Westwood Ltd.

Douglas-Westwood Associates (2004b) *The World Floating Production Report 2005–2009*. Canterbury, UK: Douglas-Westwood Ltd.

Douglas-Westwood Associates (2005a) *The World Offshore Oil and Gas Forecast 2005–2015*. Canterbury, UK: Douglas-Westwood Ltd.

Douglas-Westwood Associates (2005b) *World Deepwater Market Forecast*. Canterbury, UK: Douglas-Westwood Ltd.

Douglas-Westwood (2006) Energyfiles, January 12, 2006. http://www.energyfiles.com/information/generalfiles/pr-05-12%20Drilling.pdf (accessed April 12, 2006).

Duckers, L.J. (2006) 'Environmental Impact Assessment of Wave Energy Systems'. Paper presented at the WMTC Conference held in London, UK, 6–10 March 2006.

Eceizabarrena, Saenz Javier (2001) 'Problems and Legal Rules Regarding Fishing with Drift-Nets, with Particular Reference to European Community Law'. *Ocean Yearbook* 15: 221–245.

Electric Power Research Institute (EPRI) and Energy Innovation Input (EII) (2004) *Report on the Assessment of Offshore Wave Energy Conversion Devices*. California: E2I EPRI WP. http://www.energy.ca.gov/development/oceanenergy/E2I_EPRI_REPORT_WAVE_ENERGY.PDF (accessed August 24, 2006).

Electric Research Power Institute (EPRI) (2004a) *Offshore Wave Power in the US: Environmental Issues*. http://www.epri.com/attachments/297213_007_Wave_Envr_Issues_Rpt.pdf (accessed April 1, 2006).

Electric Research Power Institute (EPRI) (2004b) *Economic Assessment Methodology for Offshore Wave Power Plants*. http://www.epri.com/attachments/297213_002_Rev_4_Econ_Methodology_RB_12-18-04.pdf (accessed April 2, 2006).

Energy Information Administration (EIA) (1998) Natural Gas: Issues and Trends. Appendix B, Offshore Oil and Gas Recovery Technology. http://www.eia.doe.gov/pub/oil_gas/natural_gas/analysis_publications/natural_gas_1998_issues_trends/pdf (accessed October 22, 2006).

Energy Information Administration (EIA) (1999) *Natural Gas 1998: Issues and Trends*. http://www.eia.doe.gov/pub/oil_gas/natural_gas/analysis_publications/natural_gas_1998_issues_trends/pdf/it98.pdf (accessed December 21, 2005).

Eriksson, M.H., A. Moroso and A. Fiorentino (2006) 'The Vertical Axis Kobold Turbine in the Strait of Messina—a Case Study of a Full Scale Marine Current Prototype'. Paper presented at the WMTC Conference held in London, UK, 6–10 March 2006.

European Commission (EU) (2000) Green Paper 'Towards a European Strategy for the Security of Energy Supply', COM (2000), 769 final.

European Renewable Energy Research and Industry Agency (2005) *FP7 Research Priorities for the Renewable Energy Sector*. http://www.erec-renewables.org/documents/FP7/AZ_FP7priorities_010305_final.pdf (accessed May 17, 2006).

European Science Foundation (ESF) (2001). *Marine Biotechnology: A European Strategy for Marine Biotechnology*. ESF Marine Board Position Paper 4. Strasbourg: ESF.

Falkowski, P.G., R.T. Barber and V. Smetacek (1998) 'Biogeochemical Controls and Feedbacks on Ocean Primary Production'. *Science* 281: 200–206.

Field, C.B., M.J. Behrenfeld, J.T. Randerson, P. Falkowski (1998) 'Primary Production of the Biosphere: Integrating Terrestrial and Oceanic Components'. *Science* 281: 237–239.

Food and Agricultural Organization of the United Nations (FAO) (1992) *Marine Fisheries and the Law of the Sea: A Decade of Change*. FAO Fisheries Circular No. 853: Rome, Italy.

Food and Agricultural Organization of the United Nations (FAO) (1994) *Bycatch database: A global Assessment of Fisheries Bycatch and Discards*. Rome: FAO. http://www.fao.org/DOCREP/003/T4890E/T4890E00.HTM (accessed October 28, 2005).

Food and Agricultural Organization of the United Nations (FAO) (1997) *The State of World Fisheries and Aquaculture 1996*. Rome: FAO. http://www.fao.org/DOCREP/003/W3265E/W3265E00.HTM (accessed September 25, 2005).

Food and Agricultural Organization of the United Nations (FAO) (2000) *The State of World Fisheries and Aquaculture 1999*. Rome: FAO. http://www.fao.org/DOCREP/003/X8002E/X8002E00.htm (accessed September 25, 2005).

Food and Agricultural Organization of the United Nations (FAO) (2001) *Fisheries statistics: Commodities 1999*. Rome: FAO.

Food and Agricultural Organization of the United Nations (FAO) (2004) *The State of World Fisheries and Aquaculture 2004*. Rome: FAO. http://www.fao.org/DOCREP/007/y5600e/y5600e00.htm (accessed October 18, 2005).

Food and Agricultural Organization of the United Nations (FAO) (2005) *Progress in the Implementation of the Code of Conduct for Responsible Fisheries and Related International Plans of Action*, Committee on Fisheries, Rome, 7–11 March 2005.

Food and Agricultural Organization of the United Nations (FAO) /Network for Aquaculture Center Asia-Pacific (NACA) (2000) Aquaculture Development Beyond 2000: The Bangkok Declaration and Strategy. Conference on Aquaculture Development in the Third Millennium held in Bangkok, Thailand, 20–25 February 2000. http://govdocs.aquake.org/cgi/reprint/2005/115/1150240.pdf (accessed August 3, 2006).

Foustoukos, D.I. and W.E. Jr. Seyfried (2004) 'Hydrocarbons in Hydrothermal Vent Fluids: The Role of Chromium-bearing Catalysts'. *Science* 304: 1002–1005.

Fox, R. (1993) 'The Offshore Aggregate Industry in the UK'. *Underwater Technology*, 19, (2): 17–23.

Fraenkel, P. (2006) 'Marine Current Turbines Ltd's Tidal Turbine Developments: The Development of an Entirely New Energy Conversion System'. Paper presented at the WMTC Conference held in London, UK 6–10 March 2006.

Freeman, A. Myrick III (1993) *The Measurement of Environmental and Resource Values: Theory and Methods*. Washington, DC: Resources for the Future.

Furnas, C.C. (1941) 'Future Sources of Power'. *Science* 94: 425–428.

Giarini, O. (2004) 'The Riches of the Ocean for Humankind'. *Ocean Yearbook* 18: 123–136.

Glasby, G.P. (2000) 'Lessons Learned from Deep-Sea Mining'. *Science* 289: 551–553.

Glasby, G.P. (2002) 'Deep Seabed Mining: Past Failure and Future Prospects'. *Marine Georesources and Geotechnology*, 20: 161–176.

Global Wind Energy Council (GWEC) (2006) Global Wind Energy Outlook. http://www.gwec.net/fileadmin/documents/Publications/Global_Wind_Energy_Outlook_2006.pdf (accessed August 25, 2006).

Glowka, L. (1996) 'The Deepest of Ironies: Genetic Resources, Marine Scientific Research, and the Area'. *Ocean Yearbook* 12: 154–178.

Gordon, H.S. (1954) 'The economic theory of a common property resource: the fishery' *Journal of Political Economy* 62: 124–142.

Gorina-Ysern, Montserrat (2003) 'Legal Issues Raised by Profitable Biotechnology Development Through Marine Scientific Research', *American Society of International Law*. September 2003.

Griffiths G. (2005) 'Review of Oceanographic Equipment and Sensors for the Detection and Measurement of pollutants'. Southampton Oceanography Centre Research and Consultancy Report N. 99, manuscript. http://eprints.soton.ac.uk/15332/01/socrescon099.pdf (accessed September 23, 2005).

Griffiths G., N. W. Millard, S. D. Mcphail, P. Stevenson and P. G. Challenor (2001a) 'On the Reliability of the Autosub Autonomous Underwater Vehicle'. Paper first submitted to *Underwater Technology*, July 2001. http://www.soc.soton.ac.uk/OED/gxg/Autosub_Reliability_paper.pdf (accessed July 17, 2006).

Griffiths G, N. Millard, S. D. McPhail, P. Stevenson, A. S. Brierley, P. G. Fernandes, D. A. Smeed, K. Stansfield, S. A. Thorpe, T. Osborn, D. M. Farmer, P. J. Statham, D. P. Connelly, C. R. German, A. Cunningham and P. Burkill (2001b) 'Standard and Special: Sensors Used During the Autosub Science Missions programme'. Paper presented at the Symposium on Sensors and Autonomous Underwater Vehicles sponsored by the Autonomous Undersea Systems Institute, held in Miami Beach, Florida, 4 April 2001.

Grønning, T., E. Dobos, F. Dahl-Hilstad, O. A. Johansson, M. Knell and D. S. Olsen (2004) Case Study On Biotech Innovation Systems. Norway Vol. 2: Marine biotechnology. Report submitted to the OECD in the Series 'Case Studies In Innovation' (DSTI/STP/TIP(2002). Organization for Economic Co-operation and Development. http://www.forskningsradet.no/CSStorage/Flex_attachment/BiotekTIK.pdf (accessed December 20, 2005).

Group of Experts on the Scientific Aspects of Marine Environmental Protection (GESAMEP) (2001) *Planning and Management for Sustainable Coastal Aquaculture Development*. GESAMEP Reports and Studies n. 68. Rome: FAO. ftp.fao.org/docrep/fao/007/y1818e/y1818e00.pdf (accessed August 12, 2005).

Guo J., F.-C. Chiu and C.-C. Huang (2003) 'Design of a sliding mode fuzzy controller for the guidance and control of an autonomous underwater vehicle'. *Ocean Engineering* 30: 2137–2155.

Haefner, B. (2003) 'Drugs from the deep: Marine natural products as drug candidates'. *Drug Discovery Today*. Vol. 8, No. 12: 536–544.

Hannesson Rognvaldur (2004) *The Privatization of the Oceans*, Cambridge, Massachusetts: The MIT Press.

Hardin, G. (1998) 'Extentions of The Tragedy of the Commons'. *Science* 280: 682–683.

Herzig, P.M. and M.D. Hannington (1995) 'Polymetallic massive sulphides at the Modern Seafloor: a Review'. *Ore Geology Reviews* 10: 95–115.

Hook, D.J. (2006) 'Development of Unmanned Surface Vehicles'. Paper presented at the WMTC Conference held in London, UK, 6–10 March 2006. http://www.iobis.org/about/2002Berg.pdf (accessed July 30, 2005).

International Conference for the Exploration of the Sea (ICES) (2002) 6th Symposium on Acoustic in Fisheries and Aquatic Ecology held in Montpellier, France, 10–14 June 2002.

International Energy Agency (IEA) (2005a) *Annual Report 2004*. http://www.iea.org/Publications/Reports/Anrep2004/anrep2004_full.pdf (accessed May 7, 2006).

International Energy Agency (IEA) (2005b) *Key World Energy Statistics*. http://www.iea.org/dbtw-wpd/Textbase/nppdf/free/2006/key2006.pdf (accessed March 14, 2006).

International Energy Agency (IEA) (2006) Key World Energy Statistics. http://www.iea.org/dbtw-wpd/Textbase/nppdf/free/2006/key2006.pdf (accessed 25 August 2006).

International Energy Agency (IEA) and Organization for Economic Co-operation and Development (OECD) (1996) *Global Offshore Oil Prospects to 2000*. http://www.iea.org/textbase/nppdf/free/1990/oil_1996.pdf (accessed October 25, 2006).

International Energy Agency (IEA) and Ocean Energy Systems (OES) (2003) *Wave and Marine Current Energy*. Report prepared for the UK Department of Trade and Industry by Richard Boud of Future Energy Solutions, part of AEA Technology plc. DTI report number FES-R-132. AEAT report number AEAT/ENV/1054. http://www.engineering.lancs.ac.uk/REGROUPS/LUREG/REPORTS/DTI_wave_and_tidal.pdf (accessed April 24, 2006).

International Ocean Institute (IOI) (2003) *A Gateway to Sustainable Development, Proceedings of the 30th International Conference Pacem in Maribus*, held in Kiev, Ukraine, October 27–30 2003.

International Oceanographic Institute (IOI) (1980) *Seabed Mining. Scientific, Economic, Political Aspects*. Malta: IOI Occasional Paper No. 7.

International Petroleum Industry Environmental Conservation Association and International Association of Oil & Gas Producers (2002) *The Oil and Gas Industry: from Rio to Johannesburg and Beyond*. http://www.ipieca.org/downloads/WSSD.pdf (accessed April 22, 2006).

International Seabed Authority (ISA) (1999) *Deep-Seabed Polymetallic Nodule Exploration: Development of Environmental Guidelines*. Proceedings of the International Seabed Authority's Workshop held in Sanya, Hainan Island, People's Republic of China, 1–5 June 1998. Kingston, Jamaica: ISA.

International Seabed Authority (ISA) (2001a) *Global Non-Living Resources on the Extended Continental Shelf: Prospects at the Year 2000*. Kingston, Jamaica: ISA Technical Study No.1.

International Seabed Authority (ISA) (2001b) *Proposed Technologies for Deep Seabed Mining of Polymetallic Nodules*. Proceedings of the International Seabed Authority's Workshop held in Kingston, Jamaica, August 3–6, 1999. Kingston, Jamaica: ISA.

International Seabed Authority (ISA) (2002) *Polymetallic Massive Sulphides and Cobalt-Rich Ferromanganese Crusts: Status and Prospects*. Kingston, Jamaica: ISA Technical Study No. 2.

International Seabed Authority (ISA) (2004) *Proceedings of the Tenth Anniversary of the Establishment of the International Seabed Authority*, Kingstom, Jamaica, 25–26 May 2004. Kingston, Jamaica: ISA.

Irion, R. (1988) 'Instruments Cast Fresh Eyes on the Sea'. *Science* 281: 194–196.

Jimeno J., G. Faircloth, J.M. Fernández Sousa-Faro, P. Scheuer and K. Rinehart (2004) 'New Marine Derived Anticancer Therapeutics—A Journey from the Sea to Clinical Trials'. *Marine Drugs* 2: 14–29.

Kammerzell, J. (2003) 'Drilling Slump Weakens Support Vessel Market', *Offshore Magazine*. May 2003: 102–103.

Kato, N. and H.Suzuki (2006) 'Biomimetic Underwater Vehicle for Inspection of Coastal Underwater Structures'. Paper presented at the WMTC Conference held in London, UK, 6–10 March 2006.

Kemp, J. and J. O'Nians (2006) 'The OWEL Wave Energy Converter—Demonstration Phase'. Paper presented at the WMTC Conference held in London, UK, 6–10 March 2006.

Kenney J. F. (1996) 'Considerations About Recent Predictions of Impending Shortages of Petroleum Evaluated from the Perspective of Modern Petroleum Science'. *Energy World* 240: 16–18.

Kimball, Lee A. (2003) *International Ocean Governance: Using International Law and Organizations to Manage Marine Resources Sustainably*. Gland, Switzerland and Cambridge, UK: International Union for Conservation of Nature and Natural Resources.

Kingston Tivey, Margaret (1998) *The Formation of Mineral Deposits at Mid-Ocean Ridges*, Woods Hole Oceanographic Institution. http://www.whoi.edu/oceanus/viewArticle.do?id=2400 (accessed February 4, 2005).

Kornberg H. (1981) *Royal Commission on Environmental Pollution, 8th Rept*. London: H. M. Stationery Office.

Krajick, Kevin (2004) *Medicine from the Sea*, Smithsonian Institute, Smithsonian Magazine on line. http://www.smithsonianmagazine.com/issues/2004/may/medicine.php (accessed October 12, 2005).

Kvenvolden, Keith A. (1993) *Gas hydrates as a potential energy resource—A review of their methane content*. In Howell, D.G. (ed.) The Future of Energy Gases: U.S. Geological Survey Professional Paper 1570: 555–561.

Low B., R. Costanza, E. Ostrom and J. Wilson (1999) 'Human-Ecosystem Interactions: A Dynamic Integrated Model'. *Ecological Economics* 31 (20): 227–42.

Maki Habib K. (2000) 'Teleoperation and Collaborative Control'. Paper presented at the Sixth International Conference on Intelligent and Autonomous Systems (IAS-6) held in Venice, Italy, 25–27, July 2000.

Makris, N.C., P. Ratilal, D.T. Symonds, S. Jaagannathan, S. Lee and R.W. Nero (2006) 'Fish Population and Behaviour Revealed by Instantaneous Continental Shelf-Scale Imaging'. *Science* 311: 660–663.

Malakoff, D. (1997) 'Seas Yield a Bounty of Species.' *Science* 277: 487.

Malakoff, D. (2000) 'Grants Kick Off Ambitious Count of All Ocean Life.' *Science* 288: 1575–1576.

Malakoff, D. (2004) 'New Tools Reveal Treasures at Ocean Hot Spots.' *Science* 304: 1104–1105.

Mann Borgese, Elisabeth (1998) *The Oceanic Circle*. Tokyo, Japan, New York, and Paris: United Nations University Press.

May, R. (1989) 'The chaotic rhythms of life'. *New Scientist*, 124: 21–25.

Meyer, N. I. and K. Nielsen (2000) 'The Danish Wave Energy Programme Second Year Status'. Report presented at the Fourth European Wave Energy Conference held in Aalborg University, Denmark, 4–6 December 2000. http://www.waveenergy.dk/wave_forside/4english/papers/paper_aau00.pdf (accessed June 27, 2006).
Mickevich, M.F. and B.B. Collette (2000) MARBID: NOAA/NMFS's (US) Marine Biodiversity Database. *Oceanography* 13 (3): 75–78. http://www.iobis.org/about/2000Mick.pdf (accessed July 11, 2005).
Milazzo, Matteo (1998) *Subsidies in World Fisheries: A Reexamination.* World Bank Technical Paper No. 406. Washington, DC: World Bank. http://www-wds.worldbank.org/external/default/WDSContentServer/IW3P/IB/1998/04/01/000009265_3980624143705/Rendered/PDF/multi_page.pdf (accessed June 15, 2006).
Morgan, C. L. (2000) *Handbook of Marine Mineral Deposits.* In Cronan, D. S. (ed.), 145–170. Boca Raton, Florida: CRC Press.
Mukhopadhyay, R., S.D. Iyer and A.K. Ghosh (2004) 'Resource Characterization Of A New Manganese Nodule Field in the Indian Ocean'. Proceedings of the 32nd International Geological Congress held in Florence, Italy, 20–28 August, 2004. Florence: IGC.
National Oceanic and Atmospheric Administration (NOAA), Department of Commerce (2000) *Discovering Earth's Final Frontier: A US Strategy for Ocean Exploration.* Silver Spring Maryland.
National Petroleum Council (NPC) (1992) *The Potential for Natural Gas in the United States.* Washington, DC: National Petroleum Council.
Newman, D. J., G. M. Cragg and K. M. Snader (2003) 'Natural products as sources of new drugs over the period 1981–2002'. *Journal of Natural Products* 66: 1022–1037.
Normile, D. and R. A. Kerr (2003) 'A Sea Change in Ocean Drilling'. *Science* 300: 410–412.
O'Dor, R. (2003) *The Unknown Ocean: The Baseline Report of the Census of Marine Life Research Program.* Washington, DC: Consortium for Oceanographic Research and Education.
Offshore Oil and Gas Research Group (2004) A Review of Offshore Oil and Gas Development in British Columbia. A Study prepared for the Coastal First Nations by the School of Resource and Environmental Management, Simon Fraser University, May 11, 2004. http://www.livingoceans.org/oilgas/oilandgasreports/Coastal%20First%20Nations%20-%20Offshore%20Report.pdf (accessed August 11, 2006).
Organization for Economic Co-operation and Development (OECD) (1989) *Biotechnology: Economic and Wider Impacts.* Paris: OECD Publishing.
Organization for Economic Co-operation and Development (OECD) (2006) 'Energy Information Administration' *International Petroleum Monthly,* March 2006. Paris: OECD Publishing.
Organization for Economic Co-operation and Development (OECD) (1997) *Towards Sustainable Fisheries: Economic Aspects of the Management of Living Marine Resources.* Paris: OECD Publishing.
Organization of Petroleum Exporting Countries (OPEC) (2005) Annual Statistical Bulletin. OPEC. http://www.opec.org/library/Annual%20Statistical%20Bulletin/pdf/ASB2004.pdf (accessed November 1, 2006).
Orme J.A.C. and I. Masters (2006) 'Analysis and Comparison of Support Structure Concepts for Tidal Stream Turbines'. Paper presented at the WMTC Conference held in London, UK, 6–10 March 2006.

Patin, Stanislav (1999) *Environmental Impact of the Offshore Oil and Gas Industry*. New York: Eco monitor Publishing.

Patron, P. and I. Tena Ruiz (2006) 'A Smooth Simultaneous Localization and Mapping Solution to Improve Situational Awareness, Mission Planning and Re-planning for AUVs'. Paper presented at the WMTC Conference held in London, UK, 6–10 March 2006.

Patron, P., J. Evans, J. Brydon and J. Jamieson (2006) 'AUTOTRACKER Autonomous Pipeline Inspection'. Sea Trials 2005. Paper presented at the WMTC Conference held in London, UK, 6–10 March 2006.

PB Power (2006) Emerging Supply-Side Energy Technologies. Report prepared for the Ministry of Economic Development of New Zealand. Auckland, New Zealand. http://www.med.govt.nz/upload/37184/emerging-energy-technologies-report.pdf (accessed October 1, 2006).

Pearce David W. and R.K. Turner (1990) *Economics of Natural Resources and the Environment*. London: Harvester Wheatsheaf.

Pontes, M. T. and A. Falcao, (2001) 'Ocean Energies: Resources and Utilization'. World Energy Council, 18th Congress held in Buenos Aires, Argentina, 21–25 October 2001.

Rohan, Andrew D. (2000) *Chemical Investigations of Great Barrier Reef Ascidians. Natural Products and Synthetic Studies*. PhD Thesis, School of Science, Faculty of Science, Griffith University, Nathan, Queensland 4111, Australia. http://www4.gu.edu.au:8080/adt-root/uploads/approved/adtQGU2003 0102.104858/public/02Whole.pdf (accessed May 19, 2005).

Rudnick D. L., R. E. Davis, C. C. Eriksen, D. M. Fratantoni and M. J. Perry (2004) 'Underwater Gliders for Ocean Research'. *Marine Technology Society Journal*, 38 (1): 48–59.

Sanchirico, J.N. (2005) 'A Social Scientist's Perspective on the Potential Benefits of the Census of Marine Life'. *Marine Policy* 29 (6): 507–552.

Smart, J.H. (2000) 'World-wide Ocean Optics database (WOOD)'. *Oceanography*, 13–3: 70–74.

Smith, C.R., H. Kukert, R.A. Wheatcroft, P.A. Jumars and J.W. Deming, (1989) 'Vent Fauna on Whale Remains'. *Nature* 341: 27–28.

Størkersen Nils N. and Ø. Hasvold (2004) 'Power Sources for AUVs'. Paper presented at the Science and Defence Conference, held in Brest, France, 19 October 2004. http://www.mil.no/multimedia/archive/00062/Brest_Oct_2004__NJS__62326a.pdf (accessed October 2, 2006).

Sundar, V.C., A.D. Yablon, J.L. Grazul, M. Ilan and J. Aizenberg (2003) 'Fibre-optical Features of a Glass Sponge'. *Nature* 424: 899–900.

The Carbon Trust (2005) Marine Energy Challenge. Oscillating Water Column: Wave Energy Converter. Evaluation Report. http://www.carbontrust.co.uk/NR/rdonlyres/8F79783B-9969-45F9-957A-BD6A662EE688/0/owc1.pdf (accessed September 13, 2006).

The Ocean Institute of Canada (1991) *Maritime Affairs: A World Handbook*. Essex, UK: Longman Group UK Limited.

The Partnership for Observation of the Global Oceans (POGO) (2001) 'Biological Observations of the Global Ocean: Requirements and how to meet them'. Report of a Workshop Held at Dartington, Devon, UK, 28–30 June 2001. http://ioc.unesco.org/oceanteacher/oceanteacher2/ (accessed August 1, 2005).

Thiel, T. and F. Tiefsee-Umweltschutz (2001) 'Evaluation of the Environmental Consequences of Polymetallic Nodule Mining Based on the Results of the TUSCH Research Association'. *Deep Sea Research* 48: 3433–3452.

Thorpe, T. W. (1999) A Brief Review of Wave Energy ETSU-120.DTI. Department of Trade and Industry, UK.

Thorpe, T. W. (2000) 'Wave Energy for the 21st Century: Status and Prospects' *Renewable Energy World* 3 (4): 114–121.

Tietenberg, Tom (2006) *Environmental Natural Resources Economics*, London: Pearson International Edition, Addison Wesley.

Triton Consultants Ltd. (2002) Green Energy Study for British Columbia Phase 2: Mainland Tidal Current Energy. Report prepared for BC Hydro, Vancouver, Canada. http://homepage.mac.com/max.larson/Triton/download/environment3928.pdf (accessed September 10, 2006).

United Nations Educational, Scientific and Cultural Organization (UNESCO) (2001) *Information Kit UNESCO Convention on the Protection of the Underwater Cultural Heritage*. Paris, France: UNESCO. http://www.unesco.org/culture/legalprotection/water/images/infkite.pdf (accessed July 29, 2005).

United Nations Educational, Scientific and Cultural Organization (UNESCO) (2004) *Ocean Sciences Bridging the Millennia, a Spectrum of Historical Accounts*. Paris, France and Beijing, China: Intergovernmental Oceanic Commission, UNESCO and First Institute of Oceanography, China Ocean Press.

United Nations Educational, Scientific and Cultural Organization (UNESCO) (1999) *The Blue*. London, UK: EM International Ltd.

United Nations (UN) (2002) *Atlas of the Oceans*. UN-Oceans. http://oceanatlas.org/index.jsp_(accessed July 21, 2005).

United Nations Division for Ocean Affairs and the Law of the Sea, Office of Legal Affairs and International Seabed Authority (2004) *Marine Mineral Resources. Scientific Advances and Economic Perspectives*. New York: United Nations.

United Nations Environmental Program (UNEP) (2006a) *In the Front Line: Shoreline Protection and other Ecosystem Services from Mangroves and Coral Reefs*. Cambridge, UK: UNEP-WCMC. http://www.unep.org/pdf/infrontline_06.pdf (accessed May 15, 2006).

United Nations Environmental Program (UNEP) (2006b) Millennium Ecosystem Assessment 2005, *Ecosystems and Human Well-being: Synthesis*. http://www.millenniumassessment.org/proxy/document.799.aspx (accessed June 17, 2005).

United States Department of the Interior, Minerals Management Service, Offshore Minerals Management, Leasing Division, Sand and Gravel Unit (2003) Studies Within the Minerals Management Service. A Framework for Decisionmaking, Herndon, Virginia. http://www.mms.gov/SandAndGravel/PDF/MMS%20SG%20Strategic%20Studies%20Plan.pdf (accessed September 26, 2006).

Vallega, Adalberto (1992) *Sea Management. A Theoretical Approach*. London and New York: Elsevier Applied Science.

Van Dover, C.L., C.R. German, K.G. Speer, L.M. Parson and R.C. Vrijenhoek (2002) 'Evolution and Biogeography of Deep-Sea Vent and Seep Invertebrates'. *Science* 295: 1253–1257.

Van Dover, C.L., S.E. Humphris, D. Fornari, C.M. Cavanaugh, R. Collier, S.K. Goffredi, J. Hashimoto, M.D. Lilley, A.L. Reysenbach, T.M. Shank, K.L. Von Damm, A. Banta, R. M. Gallant, D. Götz, D. Green, J. Hall, T.L. Harmer, L.A. Hurtado, P. Johnson, Z.P. McKiness, C. Meredith, E. Olson, I.L. Pan, M. Turnipseed, Y. Won, C.R. Young III and R.C. Vrijenhoek (2001) 'Biogeography and Ecological Setting of Indian Ocean Hydrothermal Vents'. *Science* 294: 818–823.

Veil John A. and M. B. Dusseault (2003) Evaluation of Slurry Injection Technology for Management of Drilling Wastes. Paper prepared for the U.S. Department of Energy National Petroleum Technology Office. http://www.ead.anl.gov/pub/doc/si-tech-report-1584.pdf (accessed November 7, 2006).

Verdant Power (2005) Next Generatio/Unconventional Hydropower. Wave & Tidal Energy Devices and More, Renewable Energy Modeling Workshop on Hydroelectric Power, May 10, 2005.

Vidas, D. and W. Østreng, eds. (1999) *Order for the Oceans at the Turn of the Century.* The Hague, London and Boston: Kluwer Law International.

Wang C-H. & S-B. Wang (2005) *Theoretical development of Schaefer model and its application*, manuscript. http://www.spc.int/oceanfish/Html/WCPFC/SC1/pdf/SC1_ME_IP_3.pdf (accessed September 17, 2006).

Weber, Michael L. and Judith A. Gradwohl (1995) *The Wealth of Oceans.* NY: WW Norton & Company.

Weiss, R.F., P. Lonsdale, J.E. Lupton, A.E. Bainbridge and H. Craig (1977) 'Hydrothermal plumes in the Galapagos Rift'. *Nature* 267: 600–603.

Williams, S. J. (1986) 'Sand and Gravel Deposits within the United States Exclusive Economic Zone: Resource Assessment and Uses'. Offshore Technology Conference held in Houston, Texas, 5–8 May 1986, p. 377–384.

Wills, Jonathan (2000) *A Survey of Offshore Oilfield Drilling Wastes and Disposal Techniques to Reduce the Ecological Impact of Sea Dumping*, Sakhalin Environment Watch. 25th May 2000. http://www.offshore-environment.com/drillwastestream.html (accessed on November 12, 2006).

Wiltshire, J. (2001) 'Future Prospects for Marine Mineral Industry'. *Underwater Magazine* 13: 40–44.

World Resource Institute (WRI) (2001) *Pilot Analysis of Global Ecosystem.* Washington, DC: WRI.

World Summit on Sustainable Development (WSSD) (2002) *Integrated Management from Hilltops to Oceans.* Johannesburg, South Africa.

World Wildlife Fund (WWF) (2001) *Hard Facts, Hidden Problems: A Review of Current Data on Fishing Subsidies.* WWF Technical Paper. Washington, DC: World Wildlife Fund. http://assets.panda.org/downloads/hardfactshiddenproblemrev2.pdf (accessed July 20, 2005).

Wu M. and D. C-K Chen (2006) 'A Generic Solution to Bottom-Hole-Assembly Modeling'. Paper prepared for presentation at the 2006 SPE Annual Technical Conference and Exhibition held in San Antonio, Texas, 24–27 September 2006.

Index

Page numbers in italics refer to figures or tables.

For Product Safety Concerns and Information please contact our EU
representative GPSR@taylorandfrancis.com
Taylor & Francis Verlag GmbH, Kaufingerstraße 24, 80331 München, Germany

www.ingramcontent.com/pod-product-compliance
Ingram Content Group UK Ltd.
Pitfield, Milton Keynes, MK11 3LW, UK
UKHW022117190926
475702UK00013B/1501